Looking for Yellow-heads

Dan Gleason

Also by Dan Gleason

Birds! From the Inside Out

Looking for Yellow-heads

Written by Dan Gleason

Illustration and Design by Barbara Gleason

Photos by Dan Gleason
unless noted otherwise

CraneDance Publications
a wing of
BGleason Design & Illustration, LLC
Eugene, Oregon

Looking for Yellow-heads

Front Cover
Yellow-headed Blackbird (*Xanthocephalus xanthocephalus*)
by Dan Gleason

— • —

Library of Congress Cataloging-in-Publication Data

Gleason, Daniel W. (1948 –)

Illustration and Design by Barbara Gleason,
BGleason Design & Illustration/CraneDance Publications

ISBN 978-0-9824441-7-7

1. Birds—United States.
2. Bird watching; Birding. 3. Biology—United States

For information about reproducing any portion of this book contact the publisher:

CraneDance Publications
Barbara Gleason, PO Box 50535, Eugene OR 97405
E-mail: bgleason@bgleasondesign.com
(541) 345-3974 • www.cranedance.com

Printing History
First Edition, November 2012, 293 pages

Printed in the United States of America

D

Dedication

As a youngster, I often went fishing or hiking with my Dad. At other times, the whole family would get in the car and go for a "Sunday drive." As a result of these outings and often being outdoors, I learned to love the natural world and the creatures that inhabit it. That interest continued to grow, and it became obvious that when I entered the University of Oregon as a freshman, Biology would be my major.

It was as a Biology student at the University of Oregon that I first met Herb Wisner, then a Senior Instructor in Biology. After my freshman year, I began to help with classes as an undergraduate teaching assistant and it was there that I really got to know Herb, as we worked together. I had the great pleasure to accompany him in the field, both with his classes and privately. His love of nature, and particularly of birds, strengthened my passion for nature. Through Herb, I learned that you really could tell the difference between Black-capped and Chestnut-backed Chickadees and that they really do sound different. Being new to the world of birds, it was a thrill to realize that I, too, could easily distinguish the differences once I trained my eyes to look and my ears to listen. But even more importantly, Herb helped spark within me the idea that birding was much more than simply putting names to species. He taught everyone to look beyond identification and attempt to understand that the insect being eaten by a bird can be fascinating in its own right, as is the tree it is sitting in. Watch as a flycatcher plucks off the wings of a lacewing it just caught, before it eats it. Now look at the ground beneath you and see those neatly created dimples in the sandy soil beside your feet. Gently knock a few grains of sand into the funnel and watch the insect, hidden at the bottom, quickly snap at what it perceived to be a possible ant that has fallen into its trap. An ant lion, but more than an ant lion, it is the larvae of the lacewing that you watched being devoured by a flycatcher, and all parts of this story are fascinating. To look around, be thrilled by, and try to understand all that is a part of a bird's, and our world—that's the passion that Herb helped inspire in me and in others.

Later, I accepted a position in the Biology Department at the University of Oregon as a Lab Preparator where I helped set up, design and teach a variety of different lab courses for undergraduates in Biology. This allowed me to continue working with Herb as a colleague and good friend. In addition to his other responsibilities, he taught the Ornithology course each summer and for many years I had the privilege of helping with the class and to share some of the responsibilities. My appreciation for birds and nature became an all-consuming passion that continues to this day. When Herb retired from his duties at the University, the department asked me to continue offering the ornithology course. In 2004, I retired from my full duties, after 30 years on the Biology faculty, but I still return each summer to continue teaching that course.

So it is with much gratitude that I have the great honor to dedicate this book to my friend, Herb Wisner. His love of birds and nature was my inspiration to continue my lifelong study as well. As you read the pages ahead, I hope that some of that passion comes through and that you either have had, or will someday have a mentor, teacher and friend as I have had in Herb Wisner.

Many thanks for the direction in which you pointed me, Herb.

—Dan Gleason

Table of Contents

Dedication v
Author's Note xi
Acknowledgments xiii
Introduction xv

CHAPTER

1 A Field of Cranes 1
2 First Sightings 5
3 Those Elusive Rails 15
4 Amazing Swifts 23
5 Defense of a Hidden Nest 31
6 Who's in the Mirror? 35
7 What's on the Menu? 39
8 Petite Jewels 47
9 Masked Marauders 59
10 A Western Specialty 63
11 Cowbirds are not Villains 67
12 Saved by an Owl 73
13 Bushtits: Tiny Gray Fluff Balls 79
14 Reconsider Bird Boxes 85
15 Chickadees at the Feeder 91
16 In Praise of "Commonness" 97
17 Like Water Off a Duck's Back 101
18 Mallard: A Familiar Duck 107

19 Keeping Cool 113
20 Have you Seen a Weary Willie? 119
21 Biobigotry 129
22 What's Your Favorite Bird? 135
23 Looking for Yellow-heads 141
24 Purple Martins: America's Largest Swallow 147
25 What's in a Name? 153
26 Gobbler in the Woods 157
27 Great Blue Heron 165
28 A Green Heron Goes Fishing 173
29 Barn Owl: Evening Ghost 179
30 A Noisy, but Lovable Jay 185
31 Both Hated...and Beloved 191
32 "Tiger" of the Sky 199
33 Who's At My Feeder? 205
34 Do Birds Have a Sense of Smell? 213
35 Winter: A Time for Breeding? 217
36 Predators at your Feeder 223
37 The Importance of Vision 229
38 Angel Wing 235
39 Thistle-bird 241
40 Some Non-natives are Here to Stay 247

A

Author's Note

There is no particular relationship between the various stories in this book, so read them in whatever order you choose: from cover to cover or however piques your interest. My only hope is that you enjoy and find this book interesting, and with that interest, I also hope will come a greater appreciation of birds and concern about protecting them and caring for the environment we share with them.

Acknowlegments

The stories in this book come from a life spent watching birds, and some chapters tell of exciting or humorous events that have been a part of my birding life. Other chapters focus on life histories of particular species.

Many such chapters are adapted from my monthly column in *Home & Garden*, a monthly supplement of the *Register Guard*, the primary local newspaper in Eugene, Oregon. I wish to thank my editors Kelly Fenley and Joel Gorthy, for inviting me to submit an article for publication each month. Their support and encouragement is greatly appreciated as is that of the public who reads these columns and offers such wonderful feedback.

We also wish to express our thanks to Louise Shimmel, Founder and Executive Director of the Cascades Raptor Center, also in Eugene, Oregon, for helping us read this book one more time, with fresh (eagle) eyes, so that we can finally stop finding typos! Her in-depth knowledge and amazing dedication to the beautiful birds she has spent her life aiding is so very much appreciated by those of us who are fellow raptor-lovers.

I

Introduction

Birding is a wonderfully varied activity, satisfying the needs of everyone from the novice bird-watcher to the most ardent "lister." You can experience birding in solitude, share it with one or a few friends, or enjoy it as a group activity. However you choose to participate, birding can leave you with lasting memories.

Perhaps you remember the first time you saw a male Lazuli Bunting in his dazzling blue plumage…or maybe you have fond memories of watching wood chips fly as a Pileated Woodpecker excavated a cavity. Many birders like to search for and locate rare birds, so as to add to their "life-list," and some of us continue the search for one (or two) species that seem to always elude us (even though all of our friends are eager to tell us how many times and places they have seen the bird).

Some such memories are not of just a single bird sighting, however. Who could fail to recall with joy, and perhaps some awe, the cacophony of thousands of Snow Geese taking to the air and flying just above your head as they rise into the sky, as once occurred in the early birding experiences of my wife, Barbara in Massachusetts? Or, the quieter, but equally impressive sight of several thousand Dunlin flying in the distance. At one moment, the flock appears to be a large mass of silvery-white dots...but suddenly, they "disappear," blending into the background colors...and then, a few seconds later, they "reappear" as the flock twists and turns in near-perfect synchrony.

As time passes, you may find that your memories are not all equally pleasant and some, I suspect, that you hope will not be repeated. It is probably with some sadness that you recall the Hermit Thrush that flew into your window and died in your hand, or a goldfinch killed by a neighborhood cat. But, not all such events need to end with unpleasant consequences. There are memories of birds brought to harm that return safely to the wild because of your efforts or intervention. I once had a cat that was tormented by the Western Scrub-Jays that nested in the yard. Whenever Sam, our cat, walked near the hedge where the jay's nest was, she found herself harassed unmercifully. The jays screeched and were determined to alert the whole neighborhood that danger was near. Then, each parent took turns diving and swooping at the poor cat. I say "poor cat" because she seldom chased birds and seemed unaware that a nest of young ones was nearby. But, one day, she became very interested in the proceedings in our yard. The young jays were now out of the nest but still not quite ready to fly. Sam couldn't resist. Right before her was a blue-gray fluff-ball that squawked and fluttered, but best of all, it didn't fly away. How could any cat, even one who didn't regularly chase birds, not go after it? And so she did.

Now the adults were more frantic than ever before. I was inside and wondered what all of the commotion was. I soon had my answer. The cat grabbed the young jay in her mouth and promptly brought it into the house and dropped it at my feet. She looked up at me as if to say, "Am I not the most wonderful kitty in all the world? Look at this delightful toy I just brought you."

The young bird was somewhat dampened with cat saliva but otherwise appeared to be quite unharmed. The cat had no intention of killing it, she just wanted to show it off to me. I took the young jay back outside (making sure the cat did not follow) and placed it on the ground near its nest-mate. The parents, of course, could not distinguish that I was only trying to help and I became the target of their wrath. By the next day, though, both chicks were able to fly and were following their parents about the yard, begging for food and calling "more—MORE—MORE!!!"

Today, my wife, Barbara, and I still have cats, but we do not allow them to go outside and roam freely. Instead, they can go out onto our "catio"—a screened-in patio with a knee-wall topped by a shelf for the cats to sit on. Here they can have the feel and smells of being outside, but they cannot catch birds, fight with neighborhood cats, or risk exposure to diseases such as feline AIDS or leukemia, or to danger from other larger animals or cars.

Cats are a major predator of songbirds and even cats that you believe to be innocent probably are not. (See the end of Chapter 7 for more information about cats and their impact on birds.)

As I look back on more than forty years of bird study, I find that some of the stories I recall are more concerned with the people (including myself) and events than with the birds themselves. I remember a Christmas Bird Count in the pouring rain when the wind was a steady 40 mph with higher gusts. There was not a bird in sight and all on our team were soaked and miserable. It's the only time I have ever stopped counting before noon.

Then there was the time that I was calling for Northern Pygmy-Owls by imitating their calls, but was having no success. A short time later, we met a man coming out of the woods where I had just been calling but getting no response. He asked, "Did you see a Pygmy Owl anywhere? I've been hearing it calling but I can't find it anywhere." He was so excited. I wondered: should I tell him it was just me?

1

A Field of Cranes

Every once in a while we are fortunate enough to be present at a moment that leaves us truly in awe of the beauty of nature. The more time we can spend in the natural world, away from human urbanization, the more frequent such moments may be. One such moment stands out prominently for me.

In March of 1975, a friend and I decided to go to Malheur National Wildlife Refuge (NWR) in southeastern Oregon. This is one of the top birding sites in the West and is most popular in late May (Memorial Day weekend, especially) when most of the breeding species have arrived but there are still many migrants on the move. In March, it is still very cold and only a few of the migrants or summer breeding birds are present. The marsh, however, is filled with thousands of ducks, geese and swans.

Malheur NWR is a smaller basin within the Great Basin[1] and the water that drains into

Great Basin[1]

The Great Basin covers 184,427 square miles of the American west: almost all of Nevada, the western half of Utah, much of the eastern edge of California and most of southwestern Oregon. Mostly arid, its highest point is Mt. Whitney (14,505 ft.) in the Sierra Nevada Mountains along the western edge of the Basin and it lowest elevation is Badwater Basin (282 ft. below sea level) in Death Valley National Park. The Great Salt Lake lies along the eastern edge of the Basin and many rivers and streams lie within the Basin, but none flow out.

the region has no outlet. As a result, this is a fertile land with vast expanses of wetlands and marsh, although much of the region has been channelized and is managed. Historically, much of the management favored cattle and agriculture, but changes toward the end of the Twentieth Century lead to management decisions that benefit the area's wildlife. The water levels fluctuate from year-to-year so the wetlands available and the wildlife visible may differ from visit to visit. The lands are volcanic in origin and much of the surrounding landscape is sagebrush with pockets of juniper. Grazed by cattle since the late 1800s, much of the area continues to be grazed today.

Heading East from Eugene, we traveled through the Douglas-fir forests of the western Cascades, across the mountains and into the Ponderosa Pine forests on the east slope of the Cascades. Then we continued on to the juniper country near Bend, the last major population center before reaching Oregon's High Desert, with its sagebrush-steppe habitat. From here we would cross more than 100 miles of landscape that many people believe is simply flat and boring. (It really is not boring if you take time to stop and investigate the many marvelous things that can be found there.) After an hour-long drive, we still had many miles to go before reaching the town of Burns, the next town with any significant population. Dusk was beginning to approach when we saw a group of large birds passing overhead. They were Sandhill Cranes, coming to rest in a nearby field where there were already other cranes. We pulled to the side of the road to watch.

Sandhill Cranes, like this pair, nest at Malheur National Wildlife Refuge in southeastern Oregon

There are two varieties of Sandhill Cranes: the Greater Sandhill Crane and the Lesser Sandhill Crane. These two subspecies have different populations that often remain segregated from each other. In the High Desert country, including Malheur Wildlife Refuge, Greater Sandhill Cranes come to breed. They are relatively few in number compared with the Lesser Sandhill Cranes that continue on to Canada, and it was a flock of Lesser Sandhills that we saw fly in. These birds would continue north to their breeding grounds but as darkness approached, they were settling in to this field to rest for the night.

As we stopped the car and got out to watch, we could hear the sound of a great many cranes filling the air. In the field were perhaps 700-800 cranes, many of them trumpeting. To see and hear this many cranes at once is a great thrill. As we stood there and scanned the horizon, we could see many more birds in the air, all coming this direction. They were in small flocks of from 30 to more than 100 birds each, and there were a great many such groups dotting the sky. For the next forty minutes we continued to watch this great spectacle as wave after wave of cranes passed overhead and circled low to land in the field. It was nearly dark when the last of these great birds dropped to rest and the sound of their kin already in the field was nearly deafening by now, for we estimated that there were more than 10,000 cranes standing in this one field! If you have ever seen Sandhill Cranes then you know how magnificent these birds are. Now imagine what it is like to be close to more than 10,000 of these majestic birds. It is without doubt one of the most awe-inspiring natural events that I have ever been privileged to witness.

We still had more than 100 miles to go before arriving at our campground that night, but we didn't care how late it was. We had just seen something beyond belief and the hour of arrival was of no consequence. It had been too dark to take photos that night so even though we had gone to bed very late, we got up very early the next morning to travel back to the field in an attempt to photograph so many cranes in one place. We arrived back at the site shortly after dawn but we were too late—the field was empty—the birds had left to continue their journey northward. Although I got no photos of that spectacle, I could not be disappointed for I have a memory of something so special, that it will always be fresh and have the power to inspire me for the rest of my life.

2 

First Sightings

Does the first woodpecker you ever saw stand out clearly in your mind? Do you recall when you saw your first American Robin? What about the first warbler you could clearly recognize?

For many of us, our first bird sightings were always just "there," a part of the world in which we grew up. Only later did we learn that these were particular species and we then began to call them by their "official" names, like Downy Woodpecker or Wilson's Warbler.

Perhaps you remember when those names became important, but for many of the common species, I don't remember when I started calling them by name. I could perhaps give an approximate time when I realized that those large blue birds in our yard were Western Scrub-Jays instead of "Blue Jays" as we had always (incorrectly) called them. The

The American Robin is familiar to almost everyone. Do you remember the first time that you saw one?

recognition that Blue Jays and Scrub-Jays were two different species coincide with my growing interest in learning more about birds and how to identify them. As my identification skills increased, I realized that I already had a list of birds well-known to me, and for only a few of these could I place in my mind a "first" sighting. We may feel that the first time we see a "new" species should hold special meaning to us. This is a nice idea, but in reality that moment may not always be remembered.

Armed with field guides and a new-found interest in learning all I could about birds, I began to notice some sightings that could be acknowledged as "firsts"—but the reasons for remembering them vary, as perhaps they do for you. A trip to a new part of the country, or especially, to a new country, certainly brings with it the prospect of finding species of birds that are new to you. Once you arrive at your destination, some of your expectations may be realized but probably not all will, and a few unexpected species may turn up instead. So, do you remember the trip because of the bird species seen or do you remember the species because of the trip? Perhaps it is a particular event during the trip that brought that first encounter with a new species.

If you bird long enough, some of your first sightings will undoubtedly be the result of seeing rare birds—that is, birds not normally occurring in the area you are in or are visiting. In 1987, a Ross's Gull showed up in Newport, Oregon. The presence of this bird in the United States (outside of Alaska) was so noteworthy that birders from all over the country came to see it, so they could put it on their life-lists. It lingered from mid-February until early March, allowing many people the opportunity for an excellent view. Like many others, I also traveled to see it, although my trip was shorter than many since I live in Eugene, only about a two and one-half hour drive away. I turned off Highway 101 and drove down to the Coast Guard Station at the mouth of Yaquina Bay, stepped out of the car, looked out over the water beyond the station, and sure enough, there it was—a small gull with a black bill and wedge-shaped tail flying low over the water. This was certainly a memorable first, I think in part because it may be not just the *first* but the *only* time that I am likely to see a Ross's Gull, a bird very rarely seen outside of Siberia or the north Arctic waters over which it normally ranges.

Another "rarity" was one that I saw early in my birding career, in 1974, and I had only become a serious birder a few years earlier. To everyone's surprise, a singing male Hooded Warbler was found in a small grove of trees at a highway wayside a few miles north of Eugene, where I lived. This was the first time this species had ever been seen in Oregon so, of course, I had to make the 15 minute drive to see it. The agricultural land surrounding this wayside is planted in rye grass, so the small plot of trees and brush provided the only shelter for this bird, making it fairly easy to find with a little patience. I looked around for perhaps 10 minutes before I heard a song with which I was unfamiliar, and I followed the song, watching for any movement. Soon, out into full view, came this beautiful yellow-and-black warbler. I stayed another hour watching the bird flit about in the brush, singing all the while.

Alas, I missed good photo opportunities this time, because a few weeks earlier, while on a trip to the coast, my camera had fallen from my shoulder into the salt water of a tide pool. (As you can imagine, cameras, especially ones with electronics in them, do not survive well when immersed in salt water.) So I took no photos of this warbler but I have a lifelong memory of it that I will cherish.

I remember another first that was unusual but not as rare as the Ross's Gull or Hooded Warbler: an American Redstart. This very attractive warbler is certainly not unknown in Oregon but even now it is still an unusual bird throughout most of the state. In 1972, a male was seen at Davis Lake in the Oregon Cascades. The following year, successful breeding was reported for this species in this same location and for the next few years, these birds bred successfully there or in a nearby location. It was the male that was seen in 1972 which caught my attention. Here was a "rare" bird and not only was I seeing it but it was a very cooperative individual. I was camped at a campground a short walk from where the bird had been reported by others prior to our arrival and we were joined at the campground by a friend who brought along his mist nets and bird bands. The bird was staying in the area and was active and easily seen, so it certainly seemed like a good candidate for banding. However, this warbler apparently was not in agreement with our thinking.

Setting up the mist net was not easy. Where the redstart was most active was in a brushy, thickly-wooded area alongside Odell Creek which drains into nearby Davis Lake. The thick vegetation made it quite difficult to set up the net, but we finally succeeded in getting the net in place and then stepped back and waited...and waited... and waited. All day, the warbler sang and frequently displayed in full view. It was a marvelous opportunity to see this wonderful bird but it consistently avoided our net. The net was certainly not invisible but other birds plunged right in: Warbling Vireo, Song Sparrow, MacGillivray's Warbler and other species were all quite willing to fly into the net and receive some of our numbered "jewelry," but not the American Redstart. It dashed all around the net but never into it, and when dusk approached, we folded the net and went back to camp, hoping the next day would be more successful. Early in the morning, we awoke, ate breakfast and went down to unfurl the net for one more try. As before, a few birds were willing to test our nets and were banded and as before, the American Redstart displayed nicely before us but never flew into the net. By late afternoon, we only had a couple of hours left before it would be time to return home. Finally, the redstart neared the net. We were hopeful as it bounced from branch to branch, edging ever closer to the net. As last, we thought, an American Restart would be banded at an Oregon location. It sang, then dropped to a nearby branch just inches from the net...one more move and he would be in the net. But when he flew, it was *away* from the net! Then he turned and approached the net once more, then away and back-and-forth in a manner that seemed to be almost mocking us. Eventually, he flew right over the top of the net, not just once, but several times, over and back close to and away from the net. Then he added his final insult to our attempts—he sat right on the very top cord suspending the net between its poles! There, atop our net and only a couple of yards from where we stood, he sang his full song over and over. When he had finished (laughing at us, perhaps?), he flew off into the nearby brush and we did not see him again. He remained in this location for the rest of the breeding season, easily seen or heard by all who ventured down the trail but he never received a United States Fish and Wildlife (USFW) band as decoration around his leg. For me, this was only a minor frustration. Seeing this bird so well and so often was more than enough reward for my time.

Rare birds are certainly one way to see new birds and they do provide lasting and usually pleasant memories. But, for me, seeing a bird out of habitat is never as satisfying as seeing a bird for the first time in the habitat and location in which it belongs. My first sighting of a Black-and-White Warbler was as a "rare" bird in a park along the Willamette River here in Eugene, Oregon, but as far as I am concerned, my real first sighting of this bird was a few years later when I saw it in the mountains of North Carolina, where the bird belonged, creeping along a tree branch in its native woodland habitat. For me, this was much more rewarding and memorable than my earlier, out-of-range sighting.

J.K. (John Kirk) Townsend

(1809-1851), early American naturalist and ornithologist. In 1834, at the age of 25, he crossed the Rocky Mountains exploring the West. After over three years collecting and writing about birds and animals, he returned with many specimens, some later used by John James Audubon in preparation of his Birds of America and Viviparous Quadrupeds. Townsend's name is associated with many birds including: Townsend's Shearwater, Townsend's Solitaire, Townsend's Warbler and a few subspecies names of birds.

Another very memorable sighting was that of a Yellow-billed Cuckoo. In the early decades of the 20th Century, the Yellow-billed Cuckoo was an uncommon resident in riparian areas of western Oregon. J. K. Townsend reported it as "abundant" in the 1830s, but in the 20th Century, it was never considered to be common. By the 1930s, it seems to have disappeared from Oregon altogether. Since the 1970s, a few individuals have been seen, often years apart. So it was particularly exciting for me to see one in Oregon, where they had once occurred. We located it during a trip to Malheur National Wildlife Refuge in southeastern Oregon in 1998 where Barbara and I visited with some non-birding friends, named Carol and Randy. We

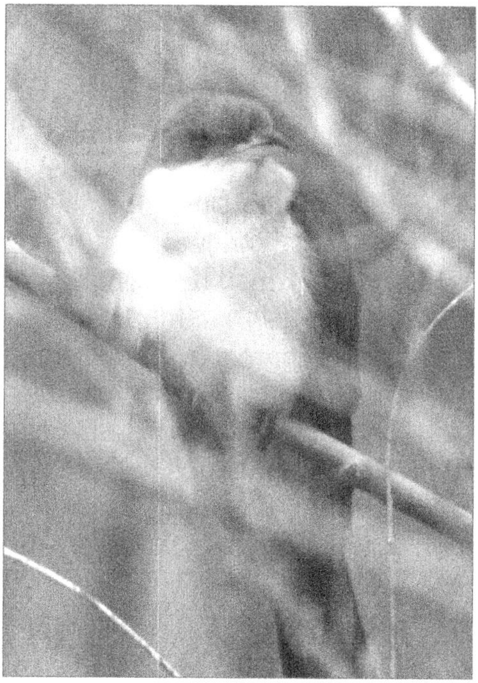

This is a photo of the first Yellow-billed Cuckoo seen in Oregon in recent times.

passed a small group of birders who told us that they had seen a Yellow-billed Cuckoo earlier, and we drove to the location where they had seen it and spent nearly an hour searching the area, but did not find the bird. We were disappointed, but eventually decided that it was time to move on.

As we started to leave, Carol pointed and said, "What's that bird?"

Now, Carol loves the outdoors but bicycling and hiking are her passions, not birding, although she tolerates her birding friends (us) and is not altogether disinterested. When Carol asks about a bird, even if it is something that most birders would consider to be very common, I always stop to look. I still enjoy looking at whatever she has seen, for even a robin may show me a behavior that I have never seen before and, thus, is always worth observing. As I looked where she was pointing, I couldn't see what she was attempting to point out, but I was probably expecting to see a Yellow Warbler or Eastern Kingbird or some other equally common bird of that area. I eventually saw a large bird, well-hidden in the willows. There it was! The Yellow-billed Cuckoo! It was my first sighting of this species, and one of the few cuckoos ever seen in recent times in this state. It was a very satisfying experience but especially because this was a bird that was once part of the normal state avifauna.

Some first sightings are purely accidental, happening when you least expect it and when you may not even be doing any active "birding." Whether I have my binoculars or not, I am always looking at and listening to any bird that is in my vicinity. I spent one memorable long weekend at Waldo Lake in the Oregon Cascades, a large, beautiful lake bordering the wilderness. There are usually not too many people there early in the season but that may be partially due to the great abundance of mosquitoes that are present until late summer. On my first full day there, I hiked the trail around the lake, a distance of over 20 miles. The second day was, therefore, merely a day of rest. I hiked only a couple of miles south of my camp and found an old campsite that was a perfect spot to sit and read and simply enjoy this warm, pleasant summer day. Three or four feet behind where I was sitting was an old fire pit and I saw Red Crossbills frequently coming down to eat some of the ashes.

Feeding on ashes is a method some birds use to obtain salt, and for crossbills, the ashes help absorb some of the sticky resin in the young conifer seeds that they eat. They also feed at clay deposits for the same reason.

I occasionally turned to watch the crossbills, but they are very common in these forests and I was enjoying my view out across the lake, my back to the fire pit and the visiting crossbills. It's always nice to see crossbills so close, so I frequently turned to observe them. As I read, I could hear them scratching in the ashes but I continued to read or look across the lake, only occasionally turning back to watch the birds. On one such occasion, I did a double-take. I had expected to see Red Crossbills, which were still there but now there were also a few White-winged Crossbills! These birds seldom wander this far south but there they were, feeding side-by-side with the Red Crossbills on the ground just a few feet from me. This was certainly a most welcome and unexpected pleasure. Not only was I seeing a species for the first time but there were several of them calmly feeding almost within arm's reach. This certainly remains a memorable, if unexpected, first sighting.

Another unexpected first sighting was that of my first Black-legged Kittiwake. These birds frequent the coastal waters of the western U.S. during the non-breeding season but are often far from shore and thus are less frequently seen than other gulls. In September 1972, while visiting a friend in San Francisco, we attended a baseball game in Oakland, accompanying her father who was a fan. I'm not much of a baseball fan but have no objection to seeing a game once in a while. This was certainly not a birding trip but our seats were in the open, near the top of the stands, and gulls and other birds occasionally flew by. I paid little attention to most of them as I was engaged in conversation and was watching the game. At one point, a gull flew within inches of our heads. How could I not look up at it? When I did look up, I could hardly believe my eyes. Circling right above my head was a Black-legged Kittiwake! It was unmistakable even without my binoculars—a small gull with yellow bill, black legs and a black triangular wingtip that contained no white spots. I had no field guide and no binoculars but I also had no doubt as to what the highlight of the evening was for me. To this day, I clearly remember that bird but I couldn't possibly tell you who won the baseball game or even, for that matter, which teams were playing.

3

Those Elusive Rails

Rails are fascinating but somewhat elusive birds, so any moment shared with one is usually memorable, whether you are a novice or an experienced birder.

One of my experiences with rails occurred on a morning trip with students from the Field Ornithology course at the University of Oregon. My good friend, Herb Wisner, was still the instructor of record for the course at that time but I participated and helped split the responsibilities. On this particular morning, we were looking for birds that inhabit the marshes along the shores of Fern Ridge Reservoir. Royal Avenue is an old roadway that becomes mostly covered with shallow water as the lake is filled each spring and summer. The roadbed that remains exposed provides some access to the surrounding marshes, making it a popular birding location.

As we walked along the roadway, Marsh Wrens hopped atop the cattails and chattered at us, their whole bodies aquiver with each song. An American Bittern's "Dunk-a-doo" sounded across the marsh and overhead a Wilson's Snipe winnowed. (In 2004, the American Ornithological Union [A.O.U.] changed this specie's name from Common Snipe to Wilson's Snipe). It promised to be a great birding morning. We told the students that we would likely hear a rail, and if we were very lucky, one might poke its head out of the marsh just long enough for us to see it.

There are only three kinds of rails found in western Oregon: Virginia Rail, Sora, and the American Coot. All are relatively common and coots are easily found on any open water, but seeing the other two rails is usually a matter of luck. On this morning, we hadn't gotten too far along on our walk before we heard the distinctive, descending whinny of a Sora. It seemed like only a minute passed before a Virginia Rail grunted in the tall marsh grasses to our left—we had our rails. With plenty of time left for the morning field trip, perhaps we might yet have a chance to actually see one.

As we approached the point where the lake's water meets the path, we stopped to look ahead across the open water that covered the old roadbed. A Great Blue Heron was standing in the distance, patiently watching for its breakfast, while an American Coot and a few Mallards swam back and forth across our view. As we watched, a Pied-billed Grebe appeared, then slowly sank beneath the water, as grebes so often do. Our patience was soon rewarded: a Virginia Rail came out of the marsh into plain view then quickly darted back into the cover of the marsh. A very brief glimpse but the students had at last seen a rail. But, an even bigger treat was yet to come.

A Virginia Rail reappeared and stood in full view in the center of the muddy road, and then, to our delight, a second rail came into view. Two rails, standing in full view at the same time! What could be better? We were about to discover an answer to that question.

Determining the sex of a rail in the field is certainly a very challenging task, but our students had no difficulty deciding the sex of these birds, for as we watched, the second bird mounted the first individual and we were treated to a demonstration of how Virginia Rails mate. The male mounted his mate four different times before both birds sought the shelter of the nearby emergent vegetation. Barely half a minute had passed when one of the birds reappeared. To our astonishment, it was followed by a little black ball of fluff attached to two enormous feet and that youngster was followed by a second fluff ball—two baby rails running across the road after the parent!! We were quick to assure the students that usually egg-laying, incubation, and hatching are time-consuming events that occur between mating and the appearance of young. In any event, it was a rare treat for any birder and certainly the highlight of this expedition.

On another occasion many years later, I was birding in the same general area with two friends. It was a pleasant, warm summer morning in late July. We had heard both Sora and Virginia Rail but had not yet seen them. My friends had never seen a rail, and as we walked along the dike we came to an area with some open water. A Pied-billed Grebe and

Fern Ridge Marsh. Dense marsh vegetation such as this can make some rails very difficult to see, although their presence is often detected by their vocalizations.

a Blue-winged Teal sat calmly on the water. Now, a Blue-winged Teal is an uncommon sight here, and was nice to see, but I was still hoping to find rails for my friends. As I looked, I saw movement in the grasses on the opposite shore. There it was: a Virginia Rail in plain view walking through the grass along the far shoreline. It was so intent on feeding along the base of the grasses that it paid us no attention, and soon it came even closer, eventually coming to only about fifteen feet away from us! Not only were my friends seeing their first rail but it was a magnificent view at close range. I was excited for them and, for myself as well. I put my binoculars down for a moment and only then did I notice that there was a second rail—standing no more than two inches from the end of my shoe was another Virginia Rail, calmly walking along the edge of the dike and paying no attention to us at all! It casually walked past us and then off into the marsh grass. I have never before, or since, had a Virginia Rail walk, almost literally, across the end of my foot. It was one of the moments to be forever cherished. (And like many such moments, a time I found myself without a camera.)

About Rails

Rails are an elusive and somewhat mysterious group of birds for many casual birders. With that in mind, I think it is perhaps useful for readers to know more about these birds.

Worldwide, there are over 140 species of rails. They belong to the same Order as cranes (Gruiformes) and are members of the Family Rallidae (Rails, Gallinules and Coots). The vast majority of these birds live in marshes, although a few prefer somewhat drier habitats. Being so closely associated with one habitat type has risks for any species. Worldwide, rails have suffered declines in populations as marshes have been channelized and drained. Floods and droughts affect habitats as do our human constructions (dams, dikes, channelization, etc.), all of which attempt to moderate these natural conditions. Pollution and other changes to water quality can also affect the suitability of habitat for rails. Small islands worldwide often have (or had) rail species endemic to that single island or to a nearby island chain.

Some experts believe that every island in the Pacific once had its own unique species of rail. Often, these birds had been left undisturbed and mostly free from predators for many centuries and many had evolved into flightless forms. Human expansion in recent centuries has threatened many of these species and most are either extinct or acutely threatened with extinction. In many cases, people changed the landscape as they settled on the islands but often, it was the non-native animals that were brought with them for which the rails had no defense or escape. These non-native invaders include: cats, dogs, snakes, goats, rats and other species that were deliberately or accidentally introduced to such islands.

In North America, Black Rail, Yellow Rail and some races of Clapper Rail are threatened, in part because of limited range and habitat availability. Black Rails are especially threatened as they tend to prefer the edges of marshlands that are often the first areas affected by changes in habitat or deterioration of habitat quality.

While coots are easy to find and observe, the majority of rails live hidden within the cover of marsh vegetation and are not easily seen or studied. They are very well-adapted to live in their marshland homes, however. When you look at the illustrations of rails in your field guide, you see birds that appear to be round-bodied and heavy, leading to a misconception that results from only seeing the image of these birds only in a profile view. If you see a rail and it turns to face you (or turns directly away), you will discover that it is not round-bodied at all but, in fact, it is very slender in this view. It is "laterally compressed" (flattened) which is where the expression "thin as a rail" comes from. A rail can also squeeze its feathers tightly against its body to further reduce its bulk and make it easy to walk between the narrow spaces created by marsh vegetation. Not only does their shape make it easier to move through the marsh, it means that a rail moves adjacent vegetation very little as it walks, thus avoiding the possibility of rustling the vegetation in a way that might attract the attention of a potential predator. In addition, most rails have long, slender toes that make it easier to walk across the layers of soft vegetation lying atop the shallow water of the marsh. (The lobed toes of coots are an exception and are designed for life on open water.) As rails walk, they place one foot directly in front of the other, minimizing obstruction to their movement through the marsh.

Rails have stout, often elongated beaks. Those with very short beaks, like the Sora, forage by pecking at food on the surface while species with longer beaks may probe into the surrounding mud in search of food. Rails have wide-ranging diets that can vary depending upon food availability and season. They eat invertebrates, some seeds and other plant materials, but are opportunistic and some have been observed eating small vertebrates (frogs, fish, snakes, eggs of other birds, etc.) as well as some occasional carrion. It is believed that rails feed mostly animal matter to their young, and most invertebrates are more plentiful during the spring and summer, thus providing a ready food source. Additionally, animal tissue provides the protein that young birds need and is more quickly digested than are plant materials.

The habitat and rather secretive nature of rails make them difficult to study and details of their natural history and behaviors are not well-known for many species. But, there are some general patterns which we presume to be true for most rails. Coots and moorhens are more easily studied and better known but their behavior may not be typical of other rails. It is thought that most species form monogamous breeding pairs that last through the breeding season and that most pairs probably have two broods per season. Polygyny (two or more females paired with a single male) has been observed occasionally in a few species but this is probably an exception to the typical pattern. The dense vegetation makes elaborate courtship displays or brightly-colored breeding patterns difficult or impossible to see at any distance so, typically, rails have dark, earth-toned color patterns that blend in with the environment. Male and female look identical, or at least, very similar. Brightly contrasting patches of feathers (often white and sometimes more prominent in the male), especially under the tail, are sometimes present and these may play a role in courtship when members of a pair, or potential pair, are in close proximity to one another. Since visibility on dense marsh vegetation is reduced, rails often have loud, distinct voices that carry easily through the marsh for vocal communication.

A healthy marsh can provide an abundance of resources for its inhabitants to feed upon, especially during the warm spring and summer months. This, combined with the dense protective vegetation, means that rails usually need to defend only a small territory within the marsh. As a result, nests may seem to be close together, almost like a loose colony. However, rails are solitary nesters and vigorously defend their own territory. Nests are typically a deep, grass-lined bowl of marsh vegetation near the edge of the water. Some species build a ramp-like entrance to the nest and some dome vegetation over the top of the nest. A few species may build floating nests. Coots usually build their nests in open water and the nest looks like it is floating but is usually firmly anchored to the bottom or to surrounding vegetation.

The number of eggs laid by rails is variable but 5-12 is the typical range. Incubation is done solely by the female in some species, but, in other species, both sexes share that responsibility. In either case, incubation begins after the first few eggs are laid. Since many rails often lay two eggs days apart, the result is asynchronous hatching and a large difference in age and size of chicks in a single brood.

Rail chicks are a comical sight to us, with their long legs and large feet attached to a small body covered with black down. In some species, the black down has streaks or spots of another color (American Coots have bright orange streaks through the predominantly black down). At a quick glance, all one sees is a tiny ball of black fluff with huge feet running through the marsh. This awkward appearance is the result of the feet becoming full-sized well before the rest of the body. Although it looks awkward and seems humorous to us, it is of great importance to young rails as it allows them to stay well above the water and the moist, sticky vegetation at the surface of the marsh. Rail chicks are semi-precocial and are thus able to leave the nest after one to three days. They usually return to the nest at night or during brooding. This nest is not always the nest that was used for egg-laying as many rails build secondary nests that may be used as nurseries within their territory. The active young follow and watch the adults and learn to begin feeding independently within just a few days of hatching.

Rails seldom are seen in flight and when they do fly, it seems brief and awkward on short, rounded wings. At first glance, they do not seem like birds well-suited to long flights. Yet, rails are often found at distances that are hundreds, or even thousands, of miles from their normal range. Some species are migratory, flying between breeding and non-breeding ranges each year. Some marshlands where rails are found are subject to large fluctuations of water and ephemeral pools that cannot provide food and cover throughout the entire year. The ability to fly to other, more favorable regions during a given season is thus an important asset to some species of rails.

For birders, rails are often best known by voice. They can be a difficult group to get to know well. Yet, they are intriguing and offer many interesting challenges for us. There are still many aspects of their lives that are not well-known to us. Take the time to learn the rails of your region and find an opportunity to study them firsthand. You will be well-rewarded with interesting observations, however brief, and perhaps you may even observe behavior that is seldom-seen or, perhaps, not yet described.

4

Amazing Swifts

Swifts amaze me each time I watch them as they dart through the air at incredible speed. Many swifts fly at speeds of 30–40 miles per hour while foraging and they are capable of even greater speed.

In western Oregon, Vaux's Swifts (pronounced *Voxes* Swifts) dart overhead on summer days, as they eat small flying insects or spiders that are ballooning through the air. (Many spiders exude long strands of silk, which act as sails, carrying the attached spider long distances through the air occasionally to high altitudes, and this is known as "ballooning.") Part of what I find so amazing is the swifts' ability to see these small insects and spiders. Large, slow-flying insects are relatively easy for me to see but small insects can be nearly impossible for our eyes to detect. When I look up to the sky or across an open expanse, I don't typically see large numbers of flying insects. If I am in a car, traveling 30–40 miles per hour, I seldom see insects at all (except large butterflies, or those unfortunate and unidentified insects who have fatal encounters with my car windshield). Yet, every day, swifts must not only see but catch large numbers of flying insects, especially during the breeding season when adult swifts must return to the nest with large quantities of insects to feed their young. Their ability to see and then make the necessary flight maneuvers to catch flying insects is truly remarkable.

William Sansom Vaux

(May 19, 1811 – May 5, 1882)

William Vaux was an American mineralogist, born in Philadelphia, and is commemorated in the name of Vaux's Swift, which was first discovered by his friend John Kirk Townsend.

Vaux's collection of mineral specimens, collected from around the world, was considered to be the finest in the United States. He was a founding member of the Academy of Natural Sciences, a member of the Zoological Society of Philadelphia, one of eight founders of the Numismatic and Antiquarian Society, and president of the Zoological Society of Philadelphia.

Four species of swifts regularly occur in North America, but there are about 100 species worldwide, comprising the Family Apodidae. Ornithologists typically break this large family into three subfamilies, all of which have at least one representative in North America.

The subfamily Cypseloidinae is mainly tropical. The Black Swift of western North America is the most northern and the only member of this subfamily to breed in the United States and southern British Columbia.

The subfamily Chaeturinae has two species in North America: the Vaux's Swift of western North America and the Chimney Swift east of the Rockies. These are probably the most familiar swifts to most readers as they are frequently seen foraging over our communities on summer days. Swifts of this subfamily are small, slender birds that appear to have almost no tail. But as is true for all birds, swifts do have a tail; and it's an interesting one to examine. The very short tail is made up of feathers whose shafts extend beyond the vanes, appearing needle-like if seen closely. These exposed shafts serve as props to help support the bird when it roosts in a chimney, cliff crevice or hollow tree. Somewhat similar tail feathers are also found in woodpeckers where they serve the same supportive function.

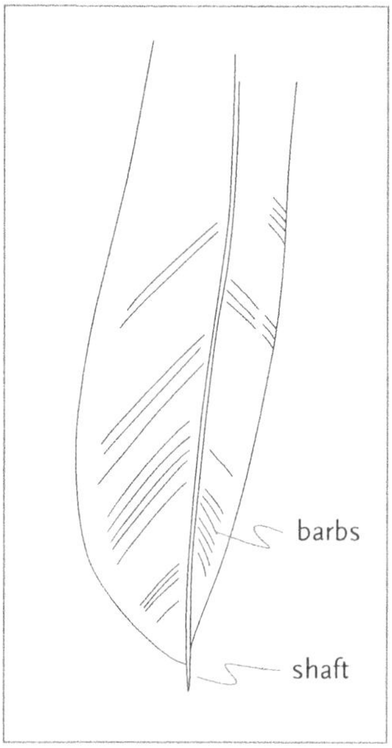

Diagram of swift tail feather, showing shaft extending beyond the barbs at the tip.

The third subfamily of swifts is called Apodinae and contains some of the world's largest swifts. The Common Swift of Europe is in this group and has a wingspan of nearly 21 inches. The White-throated Swift of the arid, American west has a wingspan of fifteen inches, and roosts and nests in cliff crevices but, these birds lack the long, extended tail feather shafts of the Chaeturine swifts. Instead, they have a unique foot that is well-adapted for clinging. The outer toes (1st and 4th digits) can freely rotate, pointing rearward, to the side, or forward, whichever position is best for grasping. It is often mistakenly said that all four toes point forward on swifts but this is true only for members of the subfamily Apodinae and only true when they rotate the outer toes forward. All other swifts have the more familiar foot pattern of three toes forward and one toe rearward.

Swifts consume large quantities of insects but feeding strategies vary between subfamilies. Black Swifts (and other members of its subfamily, Cypseloidinae) often forage long distances from their nest colonies. They may return to the nest only once or twice each day and often continue to feed until it is nearly dark. Captured insects are stored in the esophagus until the adult returns to the nest and begins the long process of regurgitation as a way of feeding the young. This feeding process may extend until well after dark. They especially seek out swarming termites, flying ants and other insects that are very rich in lipids and provide high nutritional value for the young birds.

By contrast, other swifts feed on a wider variety of insects and return to the nest several times each day, as much as once or twice each hour of daylight. They are fast fliers and may also range long distances from the nesting area. Sticky saliva holds the insects in a small, tight ball (called a "bolus") that is carried in the mouth until they return to the nest. This sticky saliva is also used to glue small twigs together, attaching them to the side of the hollow log or chimney to form the nest, a shelf of twigs upon which to lay the eggs.

Until a few years ago, I thought that swifts (at least Vaux's Swifts) only captured insects and ballooning spiders flying through the air, but I had to amend that belief after the following observation.

It was a mid-July afternoon and I was walking from the University of Oregon campus to a doctor's appointment across town. As was often the case on such pleasant summer days, I heard the rapid chatter of Vaux's Swifts overhead. These birds are common in the Eugene area during the spring and summer months and I always delight in watching their rapid, acrobatic maneuvers. One swift was rapidly circling over the buildings, nearly at roof-top level. Then the bird made a slow circle in front of me about 15 feet above the ground and briefly hovered in front of a spider's web out near the end of an oak branch. Reaching forward with its head, the bird deftly plucked the spider from its web. To my amazement, the swift circled the same tree and repeated its actions, removing another spider just eight or ten feet above the first. After capturing the second spider, it returned to the sky to join the other swifts foraging high overhead. We tend to think of swifts as catching insects at high speed but here was a case of a swift deliberately and very skillfully plucking a spider (which I had not even seen) from its web.

Each spring and fall, Vaux's Swifts of the West and Chimney Swifts in the East, put on truly amazing shows for residents of many communities. During migration, large numbers of swifts make stopovers and can congregate in large numbers. During the day, the birds spread out to forage over large areas so we are not aware of how many birds really are present. As evening approaches, they seek roosting sites. At one time, Vaux's Swifts sought old hollow snags in the forest for their roosts, but logging and other human activities have greatly reduced or eliminated such snags. Large (and hopefully, unused) chimneys now often take the snags' place. On the University of Oregon campus in Eugene there is one such large chimney. (Other similar chimneys are in Portland, Monroe, Healdsburg, CA, even downtown Los Angeles.) During the hour or so just prior to sunset, Vaux's Swifts begin to gather in the vicinity of the chimney. As nightfall approaches, the number of birds continues to grow. They begin to form tight flocks near the chimney and then they wheel and disperse again around the neighborhood, which continues for many minutes before they gather again in increasing numbers. As the light begins to wane, the swifts swirl around the chimney in tighter and tighter circles. Eventually, a few birds finally decide to "take the plunge" and drop into the chimney. More and more birds follow, forming a tight and fast spinning funnel of swifts pouring down into the chimney. During the fall migration, monitors have counted between 15,000 and 18,000 swifts and on at least one occasion, more than 20,000 birds were estimated to have entered the chimney to spend the night.

Once inside the chimney, the birds cling to the chimney's inner wall from bottom to top with the first birds to enter usually going deepest into the chimney. Once the walls are completely covered, the birds still arriving cling to the backs of those already roosting. Once settled in, they all spend the night, not leaving the chimney until well after sunrise the next day. Local residents gather each night in Eugene, Portland and elsewhere to watch this amazing spectacle and

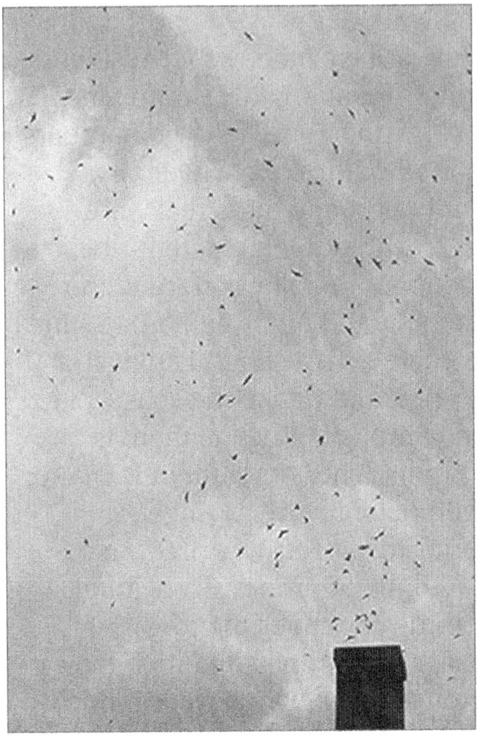

Vaux's Swifts gather in a gyre to begin their descent into large chimneys along the West coast, a phenomenon visible many spring and fall evenings.

some Audubon chapters have representatives on hand to explain what these birds are (Vaux's Swifts, not Chimney Swifts and not bats) and what is happening. It is simply one of the most amazing bird behavior sights that you can witness. When there are 15,000 or more birds, it may take as long as 45 minutes just to get all of the birds down into the chimney!

Other cities along this western migration route have this same scene played out on fall and spring nights as well, although most do not have quite as many birds go down the chimney as in Eugene. Chapman School in Portland, Oregon, is another place where equal numbers of birds may be seen and Portland Audubon also has representatives available. Both the Eugene and Portland chimneys draw large crowds of spectators each night.

In recent years, other "observers" have also come to this nightly roost but these are not human observers. Individuals of at least four different species come to prey on the swifts. In Eugene, we have seen Cooper's Hawks, Sharp-shinned Hawks, American Kestrels and American Crows all come individually at various times to snatch a swift for a late evening meal. They come on different nights and there is never more than a single predator present at any one time.

Illustration ©Barbara B Gleason, 2009

A Vaux's Swift at its nest in a chimney or tree in the Pacific Northwest.

The usual method used by all of these predators is to simply perch at the edge of the chimney and wait. As the swifts begin to funnel downward, the avian predator simply leans forward into the mass of birds and flies away with its prize. The hawks and the kestrel snatch a swift with their talons and the crow grabs its quarry in its beak. Where swifts gather to roost in other chimneys, similar predators are also attracted.

Not surprisingly, presence of one of these predators changes the behavior of the swifts. Normally, the mass of swirling birds forms a tight funnel just above the chimney. But when one of the predators arrives, the swifts seem reluctant to enter the chimney and stay outside until it is very nearly dark. Secondly, the funnel does not form in the normal fashion, at least initially. The birds congregate high above forming a large ring or giant "doughnut" high over the chimney. Instead of spiraling downward into the chimney, the birds drop individually straight down into the chimney.

It may be that they diminish their chances of being captured if they drop very fast straight past the awaiting predator. Only a very few of the numerous other swifts present begin dropping down. As the birds begin streaming by, eventually one of them is snatched by the predator. After the successful predator has flown away and been gone for several minutes, the normal funnel sometimes reforms and the birds proceed into the chimney in the usual fashion.

Pay attention to any swifts you might have in your area and see if you can discover what they do. You might one day also be amazed at what you see.

5

Defense of a Hidden Nest

Each spring, innocent pedestrians walk city sidewalks, unaware that they may soon become victims of an attack by a bird that they had no idea was nearby. Sound like Alfred Hitchcock's "The Birds" or a low-budget TV thriller? Well, it's really not as sinister as it sounds.

Every year I get calls or see a news story about people being attacked by black birds. Sometimes the avian attacker is so persistent that warnings are posted or, as happened one year in my home town, a pedestrian walkway across a major street was closed until the birds finished breeding. I myself have been the victim of these attacks numerous times, but, unlike most people, I find some enjoyment in this event.

The avian attacker is the Brewer's Blackbird. For those of you who may be unfamiliar with this bird by name, it is the robin-sized black-colored bird that often can be seen wandering

The Brewer's Blackbird is fearless in the defense of its nest.

underfoot in grocery store or shopping mall parking lots. It generally gets little attention—until spring or early summer when it sometimes makes its presence known by attacking unsuspecting passersby. Why should this gentle and usually mild-mannered bird suddenly turn against us when warm weather arrives?

In reality, the weather has little to do with this annual behavior other than indicating the arrival of spring or early summer, the breeding season for most birds, including the Brewer's Blackbird. An attack by a Brewer's Blackbird simply means that it has a nest close by, perhaps in a dense hedge along the sidewalk. The nest is well-concealed (unless you make efforts to find it) but if you are attacked, then it's because you have made the innocent mistake of walking within a few feet of the bird's hidden nest. Your presence is perceived as a threat by the birds, and, like other threats, such as potential predators, they take it upon themselves to drive away such "enemies." Both male and female blackbird may attack, but the most aggressive attack usually comes from the male. He is so bold that he may strike people on the head or shoulder. He gives little warning and often strikes even before vocalizing. The human victim, often completely unaware of the bird until they are struck, usually reacts by ducking down and looking bewilderingly around to see why and from where, she or he is under attack. Human reactions to these attacks are quite varied. Some people shriek, and run away, covering their head with a coat, purse or whatever is available. Others swat at the bird to defend themselves from further attack. A few people laugh at one small creature's attempt to drive away another creature so much larger than itself.

Thomas Mayo Brewer

(1814–1880)

Brewer was an American naturalist who collected many specimens for John James Audubon. The Brewer's Blackbird was named in his honor by Audubon. Brewer wrote the first (and only) volume of North American Oology published in 1857. The expense of producing the plates of egg illustrations for this volume was so great that no other volumes were ever issued. He may be better known as a coauthor, with Spencer Fullerton Baird and Robert Ridgway, of North American Birds (1874, in 3 volumes).

In addition to the Brewer's Blackbird, Audubon also named a duck (Brewer's Duck) and a rodent (Brewer's shrew mole) to honor Thomas Brewer. The mole is usually now referred to as Hairy-tailed mole, but the duck is somewhat of a mystery. Audubon himself thought it might have been a hybrid between Mallard and Gadwall, but some hybrids of Mallard and American Black Duck have also been suggested as possibilities.

For my part, I too am surprised when attacked, but I smile and find pleasure in this encounter and admire the boldness of this bird in protecting his mate and her eggs or young. It is this tone that I try to convey whenever I am called and asked, "What do we do about these birds that are attacking people?" I urge the callers to appreciate this small, natural drama played out in our urban environment. I want people to admire this bird and its commitment to parenthood and to realize that, while the attack is a surprise, it is seldom more than that. The bird may often physically strike you but it rarely does any damage. In the many times I have been struck by a blackbird, never has a bird so much as broken the skin. At worst, it may have misplaced a few hairs (but less than a stiff breeze would do). It certainly can be intimidating, especially to small children, but there is very little real danger. If the nest is near a busy portion of the sidewalk, we may need to redirect pedestrians for a few weeks (and maybe educate them about this event) but there is never any reason to displace the birds. Soon enough, they will have raised their young and once again be those shy and gentle birds that wander around your local parking lots.

6

Who's in the Mirror?

As breeding season approaches, male birds of many species become increasingly territorial and sometimes spend large amounts of time and energy attempting to drive away rival males. Occasionally, these actions can become misdirected into entirely useless behaviors that give the bird no benefit at all!

Each year I receive inquiries from at least one person wanting to know, "Why is this robin attacking my dining room window every morning?" or "I have this weird towhee that keeps attacking the hubcap of my car."

In each case, the bird is seeing its own reflection and attacking what it believes to be a rival male. Reflections may be second nature to us but birds don't comprehend the image as their own; they simply see a rival that won't go away.

One weekend, near the end of May, Barbara and I were just completing a wonderful trip to Malheur National Wildlife Refuge in

Like most birds, a male Mountain Bluebird will usually vigorously defend its territory from others of its kind.

southeastern Oregon. On our drive homeward, we stopped at the Sage Hen Hill rest area, just a few miles west of the town of Hines. A flash of blue across the parking lot caught my attention. It was a male Mountain Bluebird—not unusual in this area but always a delight to see.

Some bird houses have been put up on poles in this rest area and Mountain Bluebirds sometimes nest in these as well as in natural cavities in the junipers that grow nearby.

This particular male was sitting on the pavement of the parking lot when I first saw him but he then flew up and perched atop the side mirror of a large pickup truck parked in the lot. He then dropped down and hovered before the mirror for a moment then perched once again atop it. This action was repeated several times. Mountain Bluebirds often hover briefly while foraging so this didn't strike me as too unusual at first.

However, a pattern was soon evident—each time the bird hovered, it did so just a short distance in front of the truck's side mirror. Momentarily, the behavior became more aggressive. The bird would hover directly in front of the mirror and lunge forward as if to attack that other male that dared to hover in its face! It would then sit atop the mirror and rest for a moment before resuming its attack on this "rival."

Barbara grabbed the video camera and recorded several minutes of this behavior. Over the course of fifteen minutes, this male bluebird was unrelenting in its attempts to drive away this "intruder" and probably would have continued had not the owner of the pickup returned and left with his vehicle (thus removing the offending intruder).

His rival no longer in sight, the bluebird sat in a nearby juniper tree and waited for his mate who was foraging over the sagebrush on the slope just south of the rest area's lawn. There were several other cars in the parking lot with side mirrors on the doors but apparently none of them had rival males that needed to be removed.

In a somewhat more dramatic tale, once when we were visiting a nearby arboretum, we had to help intervene on the behalf of one unfortunate pickup truck driver, whose very wide, very reflective front bumper was being vigorously attacked by the resident peacock, to the point that the poor fellow wasn't able to leave the parking lot. A few of us had to try to get between the peacock and the truck to fend it off, so the fellow could turn around and make for the exit!

What's on the Menu?

As Fall approaches, there are signs of change all around us, and we feel it in the air as summer wanes. Birds tune in to these changes as well. For many species, their summer stay has already ended, or will be ending soon, and they are seeking the routes that will lead to their winter homes far to the south. But others will remain here, changing their diet or broadening the area in which they seek winter sustenance.

Through early fall, there are many wild fruits and seeds still available and even the insect-eaters still find no shortage of prey. But as the weather cools and winter begins to take hold, for many birds, food resources diminish, although they are still plentiful enough to sustain varied populations of birds that choose to winter in an area.

This is the time of year when many people think about putting out feeders to attract birds to their yards. Do the birds really need our help to find food? Probably not, but why should they

House Finches are readily attracted to sunflower seed at feeders.

turn down an easy meal? Feeding birds is really more for our benefit in connecting with nature and observing birds, but what should you offer to attract them and assure their well-being?

For birds that are strictly **insectivorous**, there is little that you can do to attract them beyond having a pesticide-free yard where insects can find plenty of cover from inclement weather.

Seed-eating birds, however, can be easily attracted to your yard. In many parts of the country, American Goldfinches, House Finches and Dark-eyed Juncos are among the easiest to attract. Chickadees, which feed mostly upon insects during the summer, change their diet and eat primarily seeds during the fall and winter. They are also among the winter residents that will readily come to a seed-filled feeder.

Many varieties of seed are now available but which is best to use? Small bags of "birdseed" are available in many stores, from variety shops to grocery stores. Much of this seed is a mixture of white millet with a small amount of larger red millet and sunflower seed. In some places in North America, white millet (also called proso millet) may attract birds, but in other places, including the Willamette Valley where we live, it is usually rejected and most of this seed simply ends up scattered on the ground, as the birds throw it aside in their search for higher-quality seed.

Sunflower seeds are one of the best and most readily-available seeds that you can use. Black oil sunflower is the best variety to use, although the striped seeds will also be eaten by some species. If you do use striped sunflower seed, be SURE that it is NOT salted!

Red-breasted Nuthatches tend to fly to a feeder, grab a seed, then fly off to consume it elsewhere. If you get a chance to watch them crack one open, you'll understand the derivation of their name: they crack it open using their beak like a hatchet.

Sunflower is the third largest vegetable oil crop in the world. Black oil seeds are highly prized by human cooks for their clear oil, preferred by Europeans for cooking, and this oil is also gaining in popularity in this country as well. The large striped variety of seed provides a good human snack when roasted, but the smaller black oil seeds provide birds with more oil, and a higher nutritional value since this oil can be converted in their bodies to excellent fuel for their high metabolisms, plus its smaller size makes it easier for small birds to open and extract the contents. This seed can be provided in a wide variety of feeders ranging from simple platforms, to hoppers with attached feeding trays at the bottom, or tube feeders specifically designed for these seeds. Smaller birds, such as chickadees, House Finches, Red-breasted Nuthatches, and goldfinches all are attracted to a tube feeder full of sunflower seeds. Larger birds, such as jays, have more difficulty sitting on the perches of these tube feeders and prefer larger bin and platform feeders.

If you are concerned about debris from the shells left behind, you can purchase sunflower seeds that have been removed from the shells. In such cases you'll not be paying for the hulls nor will it be left to be a mess or decay on the ground. This type of pre-shelled seed is for human benefit only since the birds have no difficulty in opening the shells to reach the seed.

Some birds, such as nuthatches, cache some of the seeds they find, saving them for later use. Pre-shelled seeds can dry out, losing their nutritional value and can more easily rot, especially in moist environments, so be sure to purchase freshly hulled seeds from reputable stores, since when fresh, it can be a good value purchase.

A second popular seed is Nyjer, formerly called "Niger" or "thistle." Nyjer® is actually a registered trademark of the Wild Bird Feeding Institute (WBFI). The name was registered in 1998 to avoid mispronouncing the word "niger" (which is still used in Europe) and to remove any association with thistles. Nyjer is still commonly called "thistle seed" but this is inaccurate as this plant is completely unrelated to any thistle. Many non-native thistles are noxious, invasive weeds.

The scientific name of Nyjer is *Guizotia abyssinica*. Native to Africa, this yellow-flowered member of the Aster Family has been widely cultivated in Ethiopia and India. It does not grow well anywhere in the United States, and as much as 50,000 tons of Nyjer is imported into this country from oveseas each year. The U.S. Department of Agriculture requires that all shipments of Nyjer seed admitted into the United States must be heat-treated, to prevent the germination of dodder seeds which are a common contaminant often found in Nyjer shipments. Dodder is listed as a federal noxious weed and is forbidden from being imported into this country.

The small size and very high oil content (up to 40%) of Nyjer make it a very attractive seed to some birds, especially goldfinches and Pine Siskins. It initially may seem to be an expensive seed but a little goes a long way. There are tube feeders with small openings designed just for this seed. A small mesh bag also works well as a way to provide Nyjer. The birds cling to the bag and extract the small seeds through the open mesh. If you hang such a bag, be sure to keep it covered with a dome or some other cover to protect it from the rain. The wet seed can grow mold that may be harmful to the birds.

If you want to buy some other specialty seeds, consider safflower seeds or peanuts. However, while the birds may thank you, your pocketbook may give you second thoughts about providing these in quantity. Jays and other large birds eagerly take whole peanuts and smaller birds will eat shelled peanuts. Be sure not to use salted peanuts as the excess salt may cause harm to the birds.

Another food to consider is suet. This animal fat is an excellent high-energy source of fuel for many birds. A wide variety of suet cakes are available, often with some different treats embedded in the fat. Some have sunflower seeds or peanuts and some have fruits. Despite these extra treats, it is the fat that the birds are seeking. A hanging basket of suet will often attract a flock of wandering Bushtits as well as chickadees, nuthatches, wintering warblers (such as Townsend's Warbler along the West coast) or, if you are fortunate, a woodpecker, such as Northern Flicker or Downy Woodpecker. However, be warned that if you live in a non-wooded urban area, you may find that European Starlings are also attracted to your suet. If there are starlings in your neighborhood you prefer to avoid feeding, look for a suet feeder where the suet can only be reached from below, where birds must hang upside down to access the suet from the bottom. Such feeders will often deter starlings, since they generally don't like to feed while hanging upside-down, but most of our native birds have no objection to clinging to the bottom mesh of these feeders to feast upon the suet.

And, be sure to clean your feeders regularly to keep the seed

Downy Woodpeckers are attracted to suet.

from becoming wet and moldy. Whatever food you decide to use, attracting birds can be a rewarding and enjoyable experience.

Once you have decided to provide food for wild birds, give careful consideration as to where you place your feeders so that birds will not become victims of marauding cats. All feeders need to be inaccessible to cats (many of which can jump up to six feet high) and the birds need a safe place to flee to where a cat is not waiting.

Let me take a moment to discuss the issue of cats and birds. This can sometimes be a touchy issue and I have been assailed in writing and verbally harassed when I have given public talks. I will state up front, that contrary to some people's opinion, that I do NOT hate cats! Cats are among my favorite creatures and two of them are our choice of pet. I love my cats, and they will remain indoors, which is truly a kindness to them. Indoors, they are not exposed to feline leukemia, feline AIDS, or other communicable diseases and they can't fight with stray cats, raccoons or other critters that may wander the neighborhood. I have been told that it is "natural" for cats to be outside, but domestic cats are NOT a natural part of any North American habitat. They were bred by humans and descend from wild cat in Asia, not any North American cat.

There are many misconceptions about cats and the risks and dangers they pose to birds. Many believe that their pet cat is well-fed and would never catch birds. However, even well-fed cats still have a natural instinct to hunt and they will catch birds, even if they don't choose to eat them. Declawed cats are also not safe around birds, which may surprise you. In one study of pet cats where many individual cats were monitored, the cat that caught the most birds was the one cat that was declawed. It simply leapt up and batted birds out of the air.

There is no question that cats are the cause of death of millions of birds in the United States each year, and by some estimates, more than a billion small mammals are taken by cats, to say nothing of the impact on reptiles and amphibians. This is a huge negative pressure on fragile ecosystems and also has a tremendous impact even on healthy ones. The small animals taken by cats comprise a large portion of the diet of many avian predators like hawks, accipiters and owls, so even birds not directly taken by cats can be greatly affected.

In September 2012, a study co-authored by R. W. Gerhold of the University of Tennessee's Center for Wildlife Health, Department of Forestry and Fisheries and by D. A. Jessup, a retired California Department of Fish and Game biologist, showed new findings that are of even greater concern.

Free-roaming cats (i.e., feral cats AND pets that wander outdoors freely) are an important source of animal-transmitted, serious diseases such as rabies, toxoplasmosis and plague. The estimated 125 million of these cats account for most cases of human rabies exposure among domestic animals and are the source of 1/3 of rabies post-exposure treatments in the United States. This number is likely underestimated due to inconsistent reporting. In addition,

"Trap, Neuter and Release" (TNR) programs, can lead to ***increased***, un-inoculated populations of cats that are the source of transmittable serious diseases. The study found that rabies is detected 4 times more frequently in cats than in dogs, and that rabies cases in 2010 declined in ALL domestic animals ***except*** for cats, which comprised 62% of all rabies cases in domestic animals. This is a potential hazard for many humans, both adults and children.

In particular, they found that the TNR programs so vigorously advocated in the belief that they help address the problem of the feral cat population, do ***nothing*** to help birds, continue to put the health of cats at risk and actually ***increase*** the size of feral cat colonies. This is due to a population compensation response in reaction to the in-migration of new cats to an area and the increased fertility and survival of kittens produced by in-migrating females who mate with still intact male cats in an area. (http://www.abcbirds.org/newsandreports/releases/120918.html)

I recommend visiting the American Bird Conservancy website for more information (http://www.abcbirds.org). They can provide information that has been very well-documented and researched and they have a number of on-line videos about cats outdoors, both feral and pets.

I know that there are some people who have said that scientists are simply cat-haters and are not telling the truth, but most of the scientists I know, including myself, like cats, or at the very least, wish them no harm. Many of us do love birds and the natural environment, and cats, no matter how cute or friendly, are a detriment to both. If you love birds, nature, and if you love your cats, I urge you to keep them indoors. They will live longer, healthier lives and bird and animal populations will be healthier.

We built a "catio" on the lower level of our house where there had been a partially covered patio. It is screened in so our cats can be outside yet safely enclosed to protect their health and that of the birds we also love.

8

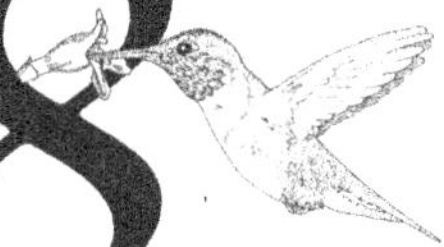

Petite Jewels

Sometimes we allow our senses to mistakenly attribute to a species qualities that do not necessarily apply. Hummingbirds are small so we might think of them as fragile or timid—but neither is true. Scientists in Colorado found one female Broad-tailed Hummingbird that was known to have lived at least twelve years, and many hummingbird species survive cold night temperatures in northern latitudes at high elevation just fine. As for being timid, I would suggest that pugnacious is a more apt term to describe many hummingbirds.

Hummingbirds sometimes seem fearless. On occasion, I have watched a Red-tailed Hawk in a treetop repeatedly being harassed by a hummingbird. Over and over the hummingbird dives within inches of the hawk until eventually the hawk has had enough and leaves. The hummingbird is in no real danger and knows it can out-maneuver and out-fly the hawk at any time.

(Also, the hawk has little interest in eating hummingbirds.) Hummingbirds are often fearless around humans as well.

Red is especially attractive to hummingbirds and many of the flowers at which they search for insects and nectar are red, but other colors can be equally attractive. I often see Rufous Hummingbirds feeding at larkspur (*Delphinium* sp.) with its tall spikes of deep blue or violet flowers, even though they are pollinated mainly by bumblebees. However, red is the color to which hummingbirds seem to be most attracted.

Many years ago, when I was camped at Davis Lake in the mountains of Central Oregon, at mid-day I was sitting at our campsite picnic table when I noticed a blur below me. There, hovering just millimeters from my red shoelaces, was a Rufous Hummingbird, coming to investigate my boots. On more than one occasion, a hummingbird has come close whenever I, or my wife, Barbara, have worn a red shirt, vest or coat. Both Rufous and Anna's Hummingbirds have flown right up and begun to feed at a feeder that is still in my hand before I can re-hang it in its usual spot. They see me as no threat so it doesn't matter whether the feeder is on its usual hook or hanging from my hand. But, if another hummingbird comes in to the same area? Well, that's a different matter, and a chase is sure to ensue!

In Oregon's Willamette Valley, there are both Rufous and Anna's Hummingbirds. The Rufous is a migrant and the Anna's is a permanent resident which has expanded its range from California over the last several decades. Both are solitary…and aggressive…little birds. Most likely you have seen photos advertising hummingbird feeders where the birds completely surround the feeder and occupy every available perch. But, you can be assured that none of the species shown are either Rufous or Anna's. Both species are very aggressive and will drive away any other hummingbird that flies near the feeder. It seems as though they expend more energy chasing one another than they do in searching for food. Often I have seen one bird chased away from a feeder by another, only to have a third hummingbird fill the void while the first two are busy chasing each other. If you have enough feeders or if they are widely enough spaced or not visible from each other, you may find more than one hummingbird feeding at the same time on different feeders. Occasionally, you do find more than one bird at a feeder, but this happens most often between mid-summer and fall and usually involves females or immature birds. Even then, their tolerance of each other is short and usually lasts less than a minute.

This sort of intolerance is obviously not found in all hummingbirds and many species happily tolerate each other and other species. The Broad-tailed Hummingbird found in Colorado's Rocky Mountains is a species that exhibits great tolerance toward other individuals, as we witnessed first-hand during a summer camping trip. Barbara brought along one of our hummingbird feeders and some sugar. On our first day in camp, she filled the feeder and hung it on a branch near the picnic table. Within a matter of minutes, a Broad-tailed Hummingbird appeared and began to feed. Within twenty minutes, each of the four feeding holes at the feeder was being used by a hummingbird and more were waiting nearby for their turn. Not only were they tolerant of each other, they had no objections to our close presence either, unless we moved too quickly. At one point, I slowly approached the feeder and stood within a few inches of it while four birds continued to feed. I raised my hand very slowly and placed it along the edge of one of the perches. Soon, one bird sat on my finger and drank from the feeder, three other birds sat at the remaining perches feeding, and one bird hovered within inches of my right ear awaiting its turn at the feeder! It was a very rewarding experience but certainly not one that is possible with the Rufous and Anna's Hummingbirds of my home state.

Hummingbirds are found only in the New World and over 330 species occur in varying habitats, ranging from Alaska to the southern portions of South America. Hummingbirds are the most numerous family of non-passerine birds in the

Female Rufous Hummingbird, at feeder

world (passerine birds are what are known as "song birds"), and among the most numerous of all bird families. (There are about the same number of species of thrushes and about 420 species of Tyrant Flycatchers found worldwide). The vast majority of hummingbird species are found in Central and South America. Within the borders of the United States there are 25 species, 16 of which are known to breed and nest north of Mexico.

At 2 ¼ inches, the Cuban, or "Bee" Hummingbird is the smallest bird in the world. Of course, not all hummingbirds are that tiny; many are larger than other birds like kinglets and warblers. High in the Andes lives the largest of all hummingbirds, the Giant Hummingbird, over eight inches long with a body larger than a Song Sparrow.

What hummingbirds lack in size they make up for by having the:

- largest brain in proportion to their size of any group of birds
- largest breast muscles in proportion to their size of any bird

- greatest rate of oxygen consumption of any bird
- greatest energy output of any warm-blooded animal
- fastest heart rate of any bird (over 1200 beats per minute)
- fastest wingbeats of any bird (over 80 beats per minute).

From banding records, we know that some hummingbirds can live at least 12 years, longer than the average life-span of many songbirds. Small species of shrews, of comparable size to hummingbirds, live only a year or less.

The Ruby-throated Hummingbird of eastern North America makes a remarkable annual journey to and from Mexico. Every fall, these birds travel from breeding sites across the eastern U.S. and Canada down to the tip of Florida. From there, they fly across the Gulf of Mexico to the Yucatan Peninsula, a journey of over 500 miles across open ocean with no place to land or find food! Prior to this long flight, each hummingbird nearly doubles its weight with stored fat, which provides fuel that is burned to power the long flight. Enough fat is stored to allow a non-stop flight of over 550 miles, which gives the birds a small, but adequate margin of error, especially since they are aided by the trade winds which are blowing east to west during the period of fall migration.

That such small birds are capable of flying such long distances may seem remarkable but it is probably the basis for some of the strange ideas that some people have about hummingbirds. Every two or three years, I get calls from people with some odd ideas about how they think hummingbirds migrate. The most frequently repeated story combines hummingbirds and geese—usually Canada Geese—and holds that hummingbirds migrate on the geese's backs. In some versions of the story, hummingbirds tuck themselves within the plumage on the back of the geese and hop off when the geese land. Another variation says that hummingbirds use their beaks to hold onto feathers on the neck or back of the goose. Some people go so far as to say that the geese actually help the hummingbirds onto their backs. Other less common versions of the story involve the use of vultures, cranes or even robins as transport for the hummingbirds! There are, of course, many reasons that these odd ideas have absolutely no truth to them. For one thing, hummingbirds, as I described above, are perfectly capable of migrating on their own without the assistance of any other species. Secondly, geese and hummingbirds migrate at different times of the year, begin from different locations and end up in different habitats!

There are also many misconceptions about how hummingbirds feed. Most people assume that flower nectar is their sole diet. Nectar certainly is a primary food, but insects, spiders and mites are also of great importance in the diet since these provide the protein, fats, vitamins and minerals necessary to sustain life. Often, a hummingbird may be observed interrupting its feeding at a flower or hummingbird feeder to fly a short distance away to catch and eat a small spider or insect that has caught its attention. Many hummingbirds also feed on tree sap dripping from small holes in tree trunks that were initially drilled by sapsuckers. In such cases, they may be taking the sap but they may also be feeding on the many small insects that are attracted to the sap.

In fact, there is a strong link between sapsucker sap-wells and hummingbirds. The Ruby-throated Hummingbird of eastern North America, actively seeks out Yellow-bellied Sapsuckers and preferentially nests in the vicinity of the sap-wells made by the sapsucker. A similar association happens in the West between Rufous Hummingbirds and Red-breasted Sapsuckers.

Hummingbirds are usually easy to attract to suburban and rural homes. Hummingbird feeders are probably the easiest way to do so but many flowers are also helpful in attracting the birds. I do NOT advise using any of the commercial "hummingbird food" mixes available to fill your feeders. These mixes frequently contain red dyes and other chemicals that are, at best, completely unnecessary, and at worst, may potentially cause harm to the hummingbirds. Even if you find a product that doesn't contain potentially harmful substances, these products are an extremely expensive way to buy simple table sugar! Although the mixes claim to contain "vitamins and minerals," these additional micronutrients are easily obtained from other natural sources on which a hummingbird feeds daily.

Common table sugar (sucrose), mixed with water, is the ONLY thing you should put into a hummingbird feeder. Use 1 part table sugar to 4 parts water; since most feeders hold about a cup or slightly more, measure ¼ c. sugar and then add 1 c. water. Stir well, then dispense into your freshly-cleaned feeder.

Sucrose is the same sugar that is found in the nectar of most flowers and is the natural food source for hummingbirds. It is the sugar most easily digested by them and it also provides the highest amount of energy. Other sugars may be difficult or even impossible for hummingbirds to digest. NEVER, under any circumstances, use honey to attract hummingbirds! Some people believe that this is a more "natural" sweetener, but it could be deadly for these birds. Hummingbirds will never naturally encounter honey as it is a product manufactured by bees and found only in their honeycomb—never in plants where hummingbirds forage. Honey is a combination of several different sugars, many of which hummingbirds cannot digest. In addition, honey may contain fungal spores that can infect the tongue of hummingbirds making them unable to feed. Honey may contribute to, or be the cause of certain liver dysfunctions in hummingbirds. In short—NEVER USE HONEY!

Hummingbirds are attracted to red, so some red color may be helpful somewhere on your feeder but it is usually not necessary. **However, NEVER put red coloring in the sugar water.** As I mentioned earlier, the red dye is really not necessary and there is some evidence that it may be harmful to the birds you are wishing to attract. Most hummingbird feeders have some red on them but even a custard dish with red nail polish around the rim will work. (Be careful, because an open container will also attract yellowjackets.) Be sure to clean your feeders routinely and change the solution frequently, especially in warm weather. Hot water is usually all that is necessary to use for cleaning, but if you do use any detergent or a weak bleach solution, be sure to rinse with clean water very, very thoroughly as any residue will be harmful to the birds. It is best to clean plastic feeders with water only, as some plastics may retain detergent residues.

A variety of flowers can also be planted to attract hummingbirds. The birds typically come to red, tubular flowers, like columbine, but they also frequent other colored flowers as well. Some common garden plants, like zinnias and dahlias, are not typically listed as hummingbird flowers but they often attract them, although not for gathering nectar, but because these flowers often harbor small insects, spiders, or mites that the hummingbirds will also eat. Like the majority of birds, hummingbirds have little or no sense of smell. Thus, flowers that have evolved to attract hummingbirds as their pollinators typically have little or no fragrance.

One of the most common misconceptions about hummingbirds concerns leaving sugar-water-filled hummingbird feeders out through fall and winter. I continue to hear people say that they have read or been told that they should remove hummingbird feeders in the fall to prevent the hummingbirds from delaying or "forgetting" to migrate. I have even seen this posted on a regional website of a state Fish and Wildlife Department, too! This may sound reasonable on the surface, but it is false and cannot be supported by ANY scientific data.

For those species of birds that migrate, there is nothing that you can do that will make them not migrate.

This instinct is so strong that you cannot change this behavior in a free-flying bird. If a hummingbird is present all year it is for reasons other than you leaving a feeder out.

A hummingbird in an area in winter may seem strange, but many species manage just fine most of the time. They are still able to find the tiny insects and mites that they need for protein (as do all hummingbirds) but they will also make regular visits to your hummingbird feeder as well.

In the Pacific Northwest, the hummingbird seen “lingering” during the fall and winter is the Anna’s Hummingbird, which is a non-migratory species and will certainly appreciate a few extra treats in the form of your hummingbird feeder. Over the last few decades, this bird has gradually been extending its range northward from California. It now can be found as far north as southern British Columbia. In the Willamette Valley of Oregon, it has become a permanent resident. During the summer it competes with Rufous Hummingbirds for access to the feeders but in the winter, it only competes with others of its own kind for feeding access. But, regardless of the season, hummingbird feeders rarely account for more than 10–15% of any one bird’s diet. Flower nectar is preferred over feeders and hummingbirds will quickly abandon a feeder to capture a small insect, spider or mite, all of which are large parts of their diet. Insect and spider eggs are also sought and eaten during the winter.

The Anna’s Hummingbird does not really need your feeder to survive winter but such feeders do become more important on the few occasions (in the Willamette Valley, at least) when temperatures drop very low and everything is frozen or covered with snow. If the contents of your feeder are actually frozen when you get up in morning, it is a good idea to make a fresh batch of unfrozen sugar-water and put it out quickly. Your hummers will certainly appreciate it. Some people have a second feeder filled and ready to go during these times. A shop-light placed above the feeder during the night may provide enough warmth to keep the feeder from freezing. If you do this, please be careful to locate it in such a manner that will not endanger the birds or create a fire hazard to your property, of course. You want to attract small birds with brilliant plumage, not large trucks with brilliant flashing red lights.

The Rufous Hummingbird of the Pacific West breeds in spring, and is migratory, spending the winter in Mexico. By fall, most Rufous Hummingbirds have already left their breeding ground, especially the males. After breeding, the vast majority move to higher elevations to feed on flower nectars, and by mid-August, Rufous Hummingbirds have each established a small feeding territory in mountain meadows. When flowers are abundant, they will feed throughout the day, converting the sugar in the nectar into fat they store in their bodies. In a week or two, they will have nearly doubled their weight and will then begin the long journey to Mexico, departing from meadows in the Rockies and the Cascades. The few that still remain in the valleys are mostly young birds.

Fat is the fuel birds use to migrate since it provides more energy than an equivalent amount of carbohydrate. In years with a poor flower supply, it becomes harder for the birds to store enough fat and they need to spend more time foraging over a wider area. The increased competition with neighboring hummingbirds expends vital energy, impacting the overall health of the birds. Many may die for lack of food prior to leaving or die from lack of enough stored reserves during migration.

From this information, we might surmise that the presence of numerous feeders left up by hummingbird-lovers in the Fall would actually ***increase*** the opportunities for birds to feed, lessen competition, and hasten, not delay, migration. So leaving your feeder up is good for the Anna's Hummingbirds that remain here, and may provide an easy source of fuel for any other hummingbirds preparing for migration.

To attract whatever hummingbirds may be found in the region in which you live, you can put up a feeder, plant hummingbird-friendly flowers or simply enjoy them where you find them on your bird outings. Wherever you do find them, spend some time watching them closely. You will be well rewarded for those few precious moments.

Masked Marauders

Someone looking up at our house from the street below could easily guess that we feed wild birds. Situated on a hill, with the main living area on the second floor, our deck with numerous bird feeders is a full story above our back yard, well above any neighborhood cats that may come through. The variety of feeders hanging from large hooks attached to the deck railing include two kinds of suet feeders, tube and hopper feeders filled with black sunflower seeds, and hummingbird feeders. All are kept filled throughout the year, including the hummingbird feeders.

About two years after moving into our home, we noticed something unusual: when we arose in the morning we'd look out the window to discover that the hummingbird feeder was completely empty. How strange! We knew it was full the night before when we went to bed. Did it have a leak? A close examination while we cleaned and refilled it revealed no explanation and no indication of a leak. We replaced it and watched the hummingbirds feed off and on throughout the day but their use did not reduce the volume significantly. Upon arising the next morning, the feeder was dry once again!

Now,I know that hummingbirds are not active during the night, even during hot summer weather and certainly not in the middle of winter. We do have owls in our area and they are active during the night but no self-respecting owl would have anything to do with a hummingbird feeder. Flying squirrels are nocturnal, but it's hard to imagine that they would be interested in sugar water. Yet, each morning the feeder would be empty and we would have to refill it. Our curiosity was heightened and we periodically looked out during the evenings to attempt to solve this mystery. One night, Barbara returned from a class and thinking about the hummingbird feeder mystery, she turned on the porch light. Ah-ha! There was the answer—Five pairs of raccoon eyes looked back at us: one adult and four youngsters.

The hummingbird feeder in question hangs from a large metal hook clamped to the deck railing, and is shaped like a shallow dish, covered by a red plastic top with four small holes through which the hummingbirds can reach the sugar-water inside. The raccoons discovered that they could stand on the deck railing, reach out and tip the feeder forward, allowing the contents to leak out beneath the cover and over the rim of the dish-shaped bottom, to spill into the raccoon's open mouth. When they weren't drinking the sugar-water, they were reaching into the holes of the nearby tube feeders and grabbing handfuls of sunflower seed.

What a treat! Sunflower seeds that are easy to obtain and a refreshing liquid dessert when you have finished! How could it get any better?

In all of my years of feeding birds, this is the first time that I have ever had raccoons drain my hummingbird feeders. The fact that the feeders hang from a second-story deck

Raccoons quickly learn that "free" food can be found at bird-feeding stations.

is no deterrent at all: the raccoons simply climb up the wooden support beams to reach the deck. To be sure that the hummers will still have food in the morning, we now hang the hummingbird feeder on the "out-of-raccoon-reach" hooks under the eaves after dark and put it back out in its proper place in the morning.

And, we also had to start doing the same for the seed-feeders, too, when several were broken by the roughness of the raccoons' enthusiastic feeding forays in the intervening years. Now all feeders are moved under the eaves at night to hooks unreachable by the raccoons.

The raccoons still come periodically to help themselves to sunflower seeds left on the deck, but we don't really mind if they take a few seeds and it's probably better for them than the sugar-water.

We recently have heard from neighbors that they are being visited by a black bear. We are only hoping the bear doesn't try to access our feeders during its nightly raids! (Although, I suspect that we would hear a bear climbing onto the deck.)

10

A Western Specialty

Since I live in the West, I like to talk and write about birds with which I am most familiar. Some, like the American Robin, can be found in much of North America, but some are strictly West Coast specialties, and the Golden-crowned Sparrow is one such species.

Before you start inviting your friends from Maryland or Ohio to come see a Golden-crowned Sparrow, keep in mind that this bird is only found south of Canada during the winter months, seen mostly in fields and hedgerows. It occurs west of the Cascades in Washington and Oregon, in portions of Central Oregon, and in western California as far south as northern Baja. However, when it is time to breed, these birds all retreat to portions of western Canada and Alaska, abandoning the lower states altogether.

When Golden-crowned Sparrows arrive in Oregon in the fall, they have a pale yellow wash on the top of the head, below which is a brown stripe above each eye. By the time they return north to their breeding grounds in mid-April to mid-May, their plumage has changed: they now sport a bright yellow crown and the once-brown face stripes have become black. They have clear gray breasts, unlike many of our other sparrows, such as Song Sparrows, which have many streaks of brown in the breast.

There is much about the Golden-crowned Sparrow that we still do not know, but one interesting fact is that this species generally migrates at night at very high altitudes. Night flocks may contain as many as 300 individuals. Some species of birds fly at high altitudes to cross mountains, but Golden-crowned Sparrows fly quite high even when the land below is at low elevation. In 1957, a Golden-crowned Sparrow collided with a small plane flying at an altitude of over 10,000 feet near San Bernardino, CA. When the plane landed, parts of the bird were recovered in the cowling of the engine allowing for its identification and providing us with some new knowledge about this species' migratory habits.

In Western Oregon, birders look forward each fall to the arrival of Golden-crowned Sparrows, but this was not always the case. In fact, early in the Twentieth Century, this bird was considered a pest and there were attempts in some states to control or eradicate the species. During the winter, it feeds only on vegetable matter and has a strong preference for cultivated grains and fruits. Not only does it eat wild grains and berries but it will take bites from apples and other commercial fruits and it will also frequent flower gardens where it eats buds, seeds and even the hearts of some flowers. Newly sprouted grasses are also a favorite food. Such food preferences did not endear these birds to grass seed farmers and commercial flower growers such as those found in Oregon where grass seed growing is a major agricultural industry. These farmers lobbied for the sparrows' removal in the early part of the Twentieth Century. Over time, we have learned that, despite these food preferences, the negative impacts of Golden-crowned

A Golden-crowned Sparrow feeds on black oil sunflower seeds on deck railing.

Sparrows are minimal, and they are no longer considered a pest species.

On their breeding grounds in the far north, adult Golden-crowned Sparrows gather insects to feed to their young and they probably eat some insects themselves. Yet, as common as they are, very little is known about their breeding habits and biology. Only one detailed study has been done and it was only an observational study of three nests, so we still have much to learn.

The song of the Golden-crowned Sparrow is well-known to many of us as a series of three plaintive notes, each on a pitch somewhat lower than the previous note. To children, it may sound like it is singing the title of a nursery rhyme, as it seems to say, "Three Blind Mice." But a more common interpretation of the song is, "Oh, Dear Me," or, "I'm So Tired," (the second syllable is a bit longer than the other two) which has earned it the nickname of *Weary Willie*. In fact, *Weary Willie* was a frequent name given to these birds by early gold miners in the Yukon who often heard their song. Many of those miners also disliked this bird because it so often was heard singing in the very places where miners found no gold, despite all of their efforts. Thus, the song became a reminder of wasted efforts and one other interpretation of the song became, "No Gold Here."

Golden-crowned Sparrows may not sing a high-spirited, bubbly song but I find the song a delight to hear on a winter day in the Willamette Valley, as soon it will again be Spring and I'll have to wait until Fall to hear the song again. Might they really be singing, "Be Back Soon"?

11

Cowbirds are Not Villains

"How do I discourage cowbirds from laying their eggs in the nest of other birds in my yard?" This question, or a variation, is sometimes asked of me. The underlying assumption is that cowbirds are villains and cause harm to our "native" birds. The truth is that cowbirds are also native to North America (I'll say more about this later) and, despite press to the contrary, they do not often cause significant damage to the populations of most other birds.

I may be in the minority, but I happen to have a special fondness for cowbirds. They, along with other brood parasites, have adopted a way of life that is quite fascinating. The process of evolution has created many very specialized species of birds and none of them should be considered villains. Evolution is a neutral process that does not create species we have any right to term "good" or "bad." It is only the human mind that imposes an arbitrary value judgment on species. But, as many of us have learned, this rush to judgment is a common problem of humans; many naturalists and writers of the past often used harsh and moralistic language, too. Here is an example of such human moralistic thinking, in the following account from American Land Birds by Alice E. Ball, 1936:

> *The four common black birds—crows, grackles, redwings, and cowbirds—all have sins laid at their door. Crows and blackbirds are grain-thieves and destroyers of the eggs and young of other birds; redwings have been accused of polygamy and theft; but if judged by human standards, none compare with cowbirds in what might be called moral degeneracy. Cowbirds not only mate promiscuously, but unlike blackbirds, have no regard for their own young. They are like the human mothers who lay their babies on doorsteps, depart, and let others rear them.*

All too often, if a species does something that we don't like, we tend to label it as bad and our bias then prevents us from seeing how beautifully adapted this species may be for survival. The Brown-headed Cowbird is often considered a "bad" bird. But set aside your bias for a moment and let's explore the way of life of a brood parasite.

I should not assume that all of my readers know about brood parasitism, so let me digress a minute to explain this behavior. An avian brood parasite is a bird that lays its eggs in the nest of another bird (most often, another species) so that the host bird will incubate and later raise the brood parasite's chick as its own. There are many different species of brood parasites and they each have somewhat different life styles. In North America, the best-known brood parasite is the Brown-headed Cowbird. In Europe, it is the Common Cuckoo. Both of these birds are what is known as an "obligate" brood parasite. That means that each of these species has no choice but to live this lifestyle; they have no other option or strategy that works for them. They never build nests and never raise their own young and have never done so. There are some other birds that exhibit brood parasitism part of the time (these are "facultative brood parasites") but do raise their own young more than half of the time. One good example is the Redhead, a duck that is a brood parasite as much as 40% of the time. It often, but not always, lays some or all of its eggs in the nest of another species of duck, frequently a Canvasback. In the case of the Brown-headed Cowbird and the Common Cuckoo, most of the host nestlings will eventually die and often only the parasite chick will survive and be fed by the host. The Canvasback will raise the parasite (Redhead) chick but will also raise most of its own chicks alongside of it. This is true of many other facultative brood parasites.

Brown-headed Cowbirds are North Americas best-known brood parasites.

The Brown-headed Cowbird was originally an inhabitant of the North American prairies. There, it followed the herds of bison where it fed on the many insects that were stirred up by the feet of the moving bison. This habit earned it the name "buffalo-bird" or cowbird. Bison are highly nomadic, seldom staying long in any one place. Through hundred of years of evolution, these birds adapted to this same nomadic lifestyle, but in so doing, nesting and raising young became impossible. It takes far too much time to build a nest, incubate eggs and raise young and still be able to forage alongside moving herds of bison. Becoming a brood parasite provided the freedom the cowbird found necessary to follow the bison herds and still find a way to assure the survival of the next generation.

Cattle Egret

The Cattle Egret in Africa has adopted the same feeding style as the Brown-headed Cowbird. It follows water buffalo and snaps up the insects disturbed by the buffalo's feet. There is usually an abundance of food and Cattle Egrets seek out by other feeding methods as well. Additionally, water buffalo are less nomadic than North American bison were. As a result, Cattle Egrets have not been forced to either abandon this lifestyle or become brood parasites and are quite capable of nest building and raising their own young.

Some brood parasites are very host-specific: they will lay their eggs in only one, or a few, species of other birds' nests. The Brown-headed Cowbird, though, is a "generalist," laying its eggs in nests of many different species of host birds. This too is an advantage for its nomadic lifestyle, as it frees the female from spending time looking for specific hosts. Instead, she can lay her eggs in any nest she finds available at the time she needs to lay an egg. A female Brown-headed Cowbird can produce up to 40 eggs each season but since many of these eggs are rejected by the intended host, this large number of eggs is a form of insurance in order to be sure that some eggs hatch and survive. When seen in this light, the cowbird is not a villain but simply has evolved a very elegant strategy that assures its survival.

Above, I stated that the Brown-headed Cowbird is native to North America but I also said that it was originally a bird of the prairie states. It is not a forest bird, although it can and does live along the margins of forests adjacent to more open lands.

We humans have dramatically changed the landscape on the continent as we have spread across it, building homes, cities, and roads, and growing crops, manufacturing products, etc. We have cleared forests and opened corridors between cleared and open lands creating clear-cuts and edge habitat in which cowbird thrive. Our activities have given the cowbird an opportunity to spread all across this continent. No longer do they need the bison or a nomadic way of life, as many of the places to which they have spread have an abundance of food, thanks to us. But the pattern of a parasitic way of life is too deeply embedded within the genes of the birds and they know no other reproductive lifestyle than that of brood parasitism. As they have spread, the number of host species whose nests they use has increased. Over 220 hosts have been reported and over 140 of those have successfully raised cowbird chicks. Many of these hosts did not co-evolve with cowbirds and have developed no defense strategy. Even so, most species have not been significantly hurt by cowbirds. Some individual pairs may be less successful in raising their own young in a particular year but the overall population is seldom hurt.

For a very few species, cowbirds may indeed have a significant impact. One such bird is the Kirtland's Warbler. Because the population of these warblers is so low, cowbirds could negatively affect the number of warblers raised in any given year. However, it was human degradation of the environment, not cowbirds that caused the warblers' decline. There is still much that we do not know, but the biggest issues for the Kirtland's Warbler is habitat destruction, both of the pine country in Michigan where it breeds and of its winter habitat in the Bahamas. The cowbird is simply a final hit that this threatened warbler cannot overcome. Again, the real villain is not the cowbird; it is humans and our destructive habits. There is an active program to search all Kirtland's Warbler nests and remove cowbird eggs to help assure the warblers' survival.

The next time that you see a cowbird, look at it with a fresh set of eyes. Evolution has freed them from the demands of parenthood, allowing a nomadic lifestyle that fills a niche left empty by other birds. Instead of rejecting these birds' habits without careful consideration, you may discover that cowbirds make fascinating subjects of study.

12

Saved by an Owl

For most of my "birding years" I have participated as a team leader in the annual Eugene, Oregon, Christmas Bird Count (or as it's commonly referred to: the CBC).

On the day of a Christmas Bird Count, I often used to get up around 4:00 AM and begin searching for owls in those pre-dawn hours. I would usually find one or two Great Horned Owls, a few Western Screech-Owls or a Northern Saw-whet Owl, as they are all relatively common in my area. Some people are good at imitating the call and drawing the owls near. I am not very good at Great Horned Owl imitations but, if persistent, I can usually locate this species anyway. Northern Saw-whet Owls and Western Screech-Owls are another matter, especially the screech-owl. I can do a reasonably good imitation of their voices and, if one is in the area, I am usually successful at getting their attention. On one occasion,

The Christmas Bird Count *is an annual winter count of each and every bird in a given count "circle," a 15 mile diameter circle in a specific locale. It has been held every year since 1900, and celebrated its 100th year in 2000. Organized by volunteers all over the world, the data is tabulated by the National Audubon Society and that data has led to many scientific understandings of the declines and falls of various species of birds, from North America to Central and South America.*

such success proved to be very fortunate for me.

On this particular Christmas Count morning, I left at my usual early hour of 4:00 AM and picked up two of my team members (others would join me later at dawn) and drove to a site that I thought would be good for finding owls. The wooded edge of the parking lot behind the local community college had often been a productive site in the past and I thought it would make a good starting point on this particular morning.

Lane Community College (LCC) is located southeast of Eugene and is just west of I-5, the main north-south interstate freeway through California, Oregon and Washington. Low, wooded hills separate the campus from the city and, at the time, the south and west sides of the campus were undeveloped woodlands. The species composition of the woods was well mixed but dominant trees included Douglas-fir and Oregon white oak. These woods are home to a rich diversity of birds and owls are certainly well-represented.

In late December, it is still quite dark at 4:00 AM and, on this particular morning, there was a cloud cover and no moonlight to assist us. Few people are up and about at that hour so it was nice and quiet for us except for some occasional traffic on the freeway in the distance. I drove to the back of the parking lot and parked under an overhead lamp illuminating the lot. The woods came to the edge of the lot and although the parking lot lamp was shining, it did not penetrate far into the woods. I decided to try my luck from that spot before walking away from the illumination to a somewhat darker area.

I whistled, imitating the "bouncing ball" sound of the Western Screech-Owl's call. No response. I tried again. Still no response. That wasn't unusual and I had expected to need to try a while before getting a response.

I was preparing to call again when we saw headlights along the roadway leading to the college that were coming in our direction. Who could it be up at this hour? Was it someone up to no good? Maybe it was another count participant coming over to see if we were owling and having any success.

But it was neither. As the car approached, we could see the bar of colored lights atop the car and the insignia on its side. It was a County Sheriff's deputy and apparently, we were "suspicious" individuals. The deputy parked beside me and got out of his car.

"'...Morning. May I ask what you're doing here at this hour?" he asked.

I explained this was the day of the Christmas Bird Count and that we were looking for owls. The look on his face told it all.

Keep in mind that today, birding is an activity in which a great many people participate and few give it a second thought. But in the 1970s, you were sometimes looked at a bit strangely if you said that you were a "bird-watcher." This deputy was obviously one of those people who thought birders were a bit odd. You could almost see the wheels spinning in his head as he tried to determine if we were engaged in suspicious activity, just nuts, or potentially dangerous nuts.

I explained about the Christmas Count and showed him our field guides and field tally sheets. He was obviously not convinced. Birding in the daylight might be okay but standing in the cold at 4:00 AM just didn't make sense.

"Why are you here at this time of morning, and why are you shining those flashlights into the woods?" he asked.

(Photo used with permission of Cascades Raptor Center.)

Western Screech-Owls are common residents of Western North American woodlands and are frequently found in wooded suburban neighborhoods.

I told him that we were just scanning the branches to see if an owl might be perched nearby and that I was just about to call for one as he drove up. It hardly seemed possible but his look of disbelief grew even stronger.

If ever I wanted to see an owl, now was the time!

I whistled my imitation. Silence, and a very suspicious look from the deputy. I whistled again. Still only silence in the woods. Before the deputy could speak, I tried my imitation a third time. A soft whinny in the woods!! This was my lucky day. Perhaps we were saved after all!

I called again and the owl answered again. I called once more and the owl flew out and sat on a nearby, overhead branch. I imitated its call and it called back.

I have never been so happy to see a screech-owl and I silently said "thank you" to that bird. It continued to call, disturbed at our presence.

The deputy just shook his head and headed back to his car. I'm not sure what he really thought of us but as he got into the seat of his car and started the engine, he grabbed his radio to report in.

His window was down and I heard him say (with disbelief in his voice), "They really are just looking for owls!"

13

Bushtits: Tiny Gray Fluff Balls

Suddenly, you see a large group of tiny birds moving through the brush. They cling to the branches in all directions, and the flock flits from branch to branch in such constant motion that you cannot even count how many birds there are. Are there only ten to fifteen, or could there be as many as forty? They are mere gray bits of fluff, twittering a constant high-pitched chatter as they move through your hedge, then on to your neighbor's yard. You have just experienced a flock of Bushtits.

Bushtits are some of our tiniest birds, only a bit longer than three inches and weighing less than 2/10 of an ounce. They are found only in western North America and range as far south as Central America. They are highly social birds at all times of the year, and flocks of 10-40 birds will remain loosely associated with each other all year long as they wander across their territory.

Bushtits that are found in Oregon are gray in color, and male and female look alike except for eye color. Males have all-dark eyes and females

Female Bushtits have white eyes, like the two birds seen in this photo.

have a light-colored iris surrounding a dark pupil.

Plumage variations are found in different geographic regions, however. In the southwest, many males have a black mask across the face and all males in Central America have the black mask. These birds were once thought to be a different species and you may find them called "Black-eared Bushtits" in some older field guides. In most field guides, you will likely find Bushtits located near the chickadees. While they do bear some resemblance to chickadees, current genetics research shows that they are probably more closely related to gnatcatchers and swallows than they are to chickadees.

Bushtits eat insects and spiders that they glean from the surface of vegetation. Most of their prey is small enough to be easily grabbed but they do sometimes take insect larvae, like caterpillars, that are too large to easily handle. When this happens, they beat the larvae against a branch until it is limp or dead and then proceed to eat it. They never eat seed that you put out, but you can usually attract them with suet cakes, especially during the winter months.

Bushtits are fascinating and were one of the first bird species ever described as having "nest helpers," non-breeding adults that assist breeding pairs with feeding and raising young. This behavior varies geographically and is much more common in southern populations and much less common in Oregon and California. Typically, over 70% of these helpers are males, either unmated or those that have lost their mate. In the areas of the southwest and in Central America where this behavior is most common, there are usually more males than females in a given population and during the breeding season, nest-helpers are the norm, rather than only the two parent birds doing the work of raising their young. Pairs without helpers are more typical of Bushtits of California and northward.

The nesting season for Bushtits lasts from mid-March through June (and is earlier in the southwest) but some nest construction can start as early as February. During this time, the flock remains in the territory but the individuals become less tightly associated with one another. Usually a single male and female from a flock will work together to build their nest and together, they defend a small territory near the nest, but less vigorously than do other kinds of birds. Once the eggs are laid, this area is no longer defended and other adults are allowed in and may even enter the nest to steal nest material for their own nest.

Nest construction begins with collection of spider webs which the birds then attach to selected branches, sometimes starting nests in several nearby locations before settling on one to complete. Once there is enough web material to support a nest structure at one site, that site is used and the nest is constructed to completion, abandoning the earlier starts. Both male and female continue to work on the nest using spider silk and soft plant materials, and the final nest is a marvel of avian construction. Tightly woven, it hangs

Bushtit nest; hole is towards the top, hidden by leaves in this photo.

like a silken sock or gourd from the branch, with the entrance hole on the side near the top where the nest is attached to the branch. It looks much like a soft slipper hanging by its heel. Four to eight eggs (six is average) are laid in the bottom of the nest and both parents share incubation duties for 11-12 days before the eggs begin to hatch. The young are ready to fly after about 18 days after hatching, when they will then leave the nest. These same parents will often raise a second brood in the same nest, but even if another brood is started, the adults still faithfully feed their first youngsters for at least two more weeks. During this period, you may see the young huddled together on a branch waiting to be fed. Such gatherings may include fledglings from other nests that are attracted to the newly emerged young.

As nesting season comes to an end, the flock begins to gather together more closely to continue their group feeding forays through neighborhoods. If you are lucky enough to find Bushtits nesting nearby, then watch for them to do so again next year. Although they will not usually reuse the same nest

Bushtits are easily attracted to a suet feeder, especially during their non-breeding season.

from year to year, they will nest in the same general vicinity.

These are delightful birds to watch. If you have a suet feeder that they have discovered, you will see one or two birds come to feed from it. Moments later, the whole flock will suddenly descend upon your feeder, taking small bits of suet. As quickly as they appear, they will move on their way to find other foraging sites within their territory. But don't despair, for they will very likely return within the hour to feed again.

14

Reconsider Bird Boxes

It's Spring, the weather is warm, birds are all around and you need a new project. Are you thinking that a bird box would be a good project? Building a bird box is a popular idea but what are the benefits and drawbacks of a box to your potential new avian residents?

When thinking of putting up a bird box, your first two questions should be: Who will the box serve? Where do you plan to put it? Many people who put up bird boxes are under the impression that almost any bird will use one. In fact, most birds in suburban neighborhoods have no interest whatsoever in a box, because they are not cavity-nesters. Only cavity nesting birds will ever use a nest box.

Robins, towhees, finches, song sparrows and most other backyard birds will not use a nest box, as their nests are open cups of twigs, mud and other plant materials. Chickadees and nuthatches are cavity-nesters but usually prefer

A House Wren leaving its nest box to seek food for the young birds within.

to select their own location and excavate their own cavity. At times, chickadees can be enticed to use a box. If swallows are in your neighborhood and you place the box properly, they *might* select your box, especially Violet-green Swallows, for those of you who live in the west.

Unfortunately, in most residential neighborhoods, you are most likely to attract European Starlings and House Sparrows, both of which are aggressive, non-native birds. In town, and to a lesser extent, in rural areas, House Sparrows will usually drive out the native swallows, taking boxes for themselves. Starlings are even more aggressive and will displace much larger birds such as woodpeckers (especially Northern Flickers) and even Wood Ducks. European Starlings and other non-native birds should not be encouraged as they can have a negative impact on nesting success of many of our native cavity-nesting birds. Starlings have caused problems for many populations of bluebirds, especially the Eastern Bluebird. Many bluebird populations initially have suffered due to habitat loss and the Starling is just one more negative impact.

A hole smaller than 1 ½ inches in size will keep Starlings from entering a box but it will also keep out many of the native birds you may want to attract. However, House Sparrows are not excluded until a hole is smaller than 1 ¼ inches in diameter, which may exclude many native birds as well.

You may find plans for boxes to attract House Wrens or Purple Martins but neither will come to suburban yards in western North America. The western race of Purple Martins rarely uses a box and is not seen in urban landscapes. House Wrens are attracted to boxes in eastern North America but seldom come to urban areas in the west, where they are common in rural landscapes. They may come to a box if it is not too near your dwelling, but they too, can have a negative impact on local birds, and a box that you put up for Western Bluebirds may be used only by House Wrens.

House Wrens, though native, do have some habits which we humans may disapprove. A male House Wren will find as many cavities within his breeding territory as he can, including your boxes, and he will proceed to fill them with sticks so that no one else may use them. His strategy is effective because it excludes others from nesting, and it also helps attract a mate while setting his territorial boundaries. The swallows and bluebirds you wanted to attract may have no boxes to use, despite your desires, but the worst is yet to come. After the wren has established his territory and secured a mate, he will locate nests of other birds that have nested nearby and poke holes in their eggs. This way, he helps assure that his young will have a better chance of survival by eliminating potential competitors for nearby food. It can also mean a near or even total failure of swallow and bluebird nests on your rural property. Although this frustrates your efforts, it is a natural behavior and it is illegal to do anything to harm or disrupt wrens, or any other native species of bird, as they are all protected by Federal law.

European Starlings and House Sparrows are non-native, however, and therefore not protected by the laws that protect native birds. Many people will encourage you to destroy their nests, eggs or even the birds themselves, but this is not something most of us are willing or able to do. It is easier to avoid doing things that invite these birds into your yard.

There is more to consider than just competing birds. You need to place the box in such a manner that the birds have easy and safe access. Be sure the box is high enough to prevent wandering cats from reaching it. Consider placing the box on a metal pole. This will make it more difficult for predators such as snakes or raccoons to reach it. A large inverted cone on the pole or some other device may be needed to keep resourceful raccoons out. They are persistent in attempting to reach a box, for they know that there is likely something good to eat inside. Even squirrels can be problematic by enlarging holes in boxes to gain entrance and use the box for their own nest. Squirrels may even eat the eggs if they can gain access into an occupied box.

If, after carefully considering all of the potential risks, you decide that you can responsibly maintain a nest box, what is your next consideration? Realize that this is a long-term responsibility and not a one-time event. Boxes must be made of materials that are safe for the birds, such as untreated wood. Metal boxes can build up considerable heat in the sun and this heat can kill young birds inside. A box must have proper ventilation to allow some natural cooling and it must be constructed so that you can easily open the box to clean it after the nesting season is complete. A box that is nailed shut is a poor design.

Now, assuming that you succeed in attracting native birds to your box, your job is still not done. After the young birds have left the nest box, you have some cleaning to do. Leaving the old nest in place may discourage its use in succeeding years and old nests can harbor parasites harmful to birds that may use the box next year. Often, it is recommended that you clean out the old nest material in the fall, but it may be better to wait until early March, after the coolest weather but before the birds return to the box. Blowflies are a common parasite on young birds and while the adult flies present no problem, the larvae attach themselves to the feet and fleshy parts around the eyes or beak of nestlings and weaken or even kill them. There is a tiny parasitic wasp that attacks blowflies and helps control their population. By waiting until early spring to clean your box, you allow the blowflies a greater chance to survive but you also allow more of their parasites, the wasps, to survive to help control the new population of flies.

There is another consideration to keep in mind when you plan to put up a box. How much enjoyment will you derive from it? You will not see any activity within the box and may only be able to see the adults come and go with food. If you are lucky, you may get to see the young ones leave but once they are out of the box, they will not return.

If you have the opportunity to leave any snags in your yard, it may provide natural cavities or places where some birds can excavate their own cavity. But I often simply encourage people to build a bird feeder instead. Many birds will use it and you will have the pleasure of seeing birds throughout the year. You also will not be encouraging starlings or House Sparrows to proliferate. By putting up a feeder, you will have done more good for more birds and for yourself than any single nest box will do.

15

Chickadees at the Feeder

What could be more delightful than having chickadees around your home throughout the day? These petite birds seem to be constantly on the go. For those of us who live in the northern portion of North America, the Black-capped Chickadee is probably the most familiar. Named, in part, for the black on the top of its head, it also has a black throat, and there is a broad triangular patch of white between the cap and throat on each side of the face. The name "chickadee" is a reference to the call, a buzzy chickadee-dee-dee. *I say "call" because that familiar voice saying* chickadee-dee-dee *is not the "song" at all, but rather a "call" that is done by both male and female.*

In the bird world, "song" is loosely defined as a male vocalization (with a few exceptions) that advertises availability for mating or proclaims territorial rights. "Calls" are vocalizations done by both male and female for a variety of reasons such as communication, keeping a flock together or sounding an alarm. The true song of the male Black-capped Chickadee is a series of 3-5 clear whistled notes. Donald Kroodsma, author of *The Singing Life of Birds: The Art and Science of*

Song *is a very complex subject and if you wish to explore some of its subtleties more closely, I would strongly recommend reading Donald Kroodsma's delightful book* The Singing Life of Birds. *This, or another of the many available books on bird song, will help improve your understanding and appreciation of birds.*

Listening to Birdsong, says that the song sounds like *Hey-sweetie*. This interpretation works for chickadees all across the country except in two places: Martha's Vineyard in Massachusetts and chickadees found west of the Cascades. In those two places, Kroodsma notes that there are subtle variations of this song that may be heard. Living here in the Pacific Northwest, I am more familiar with the western song variant than I am with most of the Black-capped Chickadees in North America.

Having lived in western Oregon, I have been able to become well-acquainted with three species of chickadees: Black-capped Chickadee, Chestnut-backed Chickadee and Mountain Chickadee. I live in the Willamette Valley, which is at a low enough elevation that the Mountain Chickadee is seldom seen here, even in the most severe winters. However, they are not at all uncommon at higher elevations in the Cascades Mountains east of the Willamette Valley. The Chestnut-backed Chickadee is the most common chickadee in coastal coniferous forests and it is also common in coniferous woodlands inland as well. I am fortunate that both Black-capped and Chestnut-backed Chickadees visit my yard nearly every day. Both species love the black oil sunflower seeds that are in our many feeders and both also frequent the suet blocks that we provide. They come at all seasons and are an endless joy to observe. They are somewhat less visible, but still present, in the yard during the summer months when they are nesting. In the spring, we have sometimes put tufts of cat hair into an empty suet basket. (We have a long-haired cat that lives indoors but sheds large amounts of luxurious long fur.) We then watch the chickadees gather great mouthfuls of this fur to take back to their nest cavity to use as a soft lining material in the nest. It is amusing to see a chickadee carrying

Black-capped Chickadees are usually easily attracted to back yard bird feeders and are especially fond of black-oil sunflower seeds.

a large ball of cat hair larger than its head! (How can it see where it is going, we wonder?)

I have had people tell me that they only have chickadees in the winter, as they assumed the chickadees had migrated to another region to breed. But chickadees are mostly a non-migratory species and if you look closely in your neighborhood, you will probably still find them present in any season. The reason that they may seem less common in the summer has to do more with their behavior and physiology rather than any geographic movements.

In the winter, chickadees depend more heavily on seed for food and so are seen frequently taking black-oil sunflower seeds from your feeder. (Keep your feeder full of these seeds: chickadees, and many other birds, love them.) During the winter, their gizzard becomes enlarged and well-developed. (The gizzard is the muscular rear portion of a bird's stomach that can thoroughly grind seeds or other hard materials.) As spring approaches, changes in structure and physiology of the digestive system occur. The gizzard degenerates to a degree and the proventriculus, the front, glandular part of the stomach, develops and enlarges. This means that during spring and summer, seeds are more difficult to pulverize and digest, and insects, now increasingly abundant, are much more easily digested.

The spring and summer seasons also coincide with the breeding season. Once the young have hatched, the adults will spend much of their time searching for insects to feed those hungry mouths awaiting their meal in the nest. Baby chickadees (and most young birds) are not able to digest seeds yet, so insects provide the protein and other essential nutrients necessary to survive and grow.

So, rather than come to your feeder for seed, they mostly glean insects from the surfaces of trees and shrubs. Keep in mind that there are few absolutes in nature and you might occasionally see a chickadee taking seed in the summer, but it is with much less frequency that they do so.

There are other reasons that chickadees can be more easily seen in winter. During the late fall and winter, they are less hidden by foliage as the leaves fall, they forage over a larger home range in search of seeds and they are much less secretive during these non-breeding periods. So, take a moment to look and listen closely, at all seasons, and you will discover that chickadees did not "arrive for the winter," but are actually present all year long. And while you are looking for chickadees, don't forget to look for many other natural wonders that are, perhaps, only an arm's length away.

The Stomach of Birds

In birds, the stomach is a two-chambered structure. The forward portion of the stomach, known as the proventriculus, is a glandular organ and the rear portion of the stomach is called the gizzard, and is thick-walled and muscular. The most developed proventriculi are found in fish-eating birds and raptors. It is in the proventriculus where secretions of highly acidic (pH 0.2–1.2) peptic enzymes and hydrochloric acid are produced. These aid the bird in its ability to digest bones and flesh.

The gizzard, or muscular portion of the stomach, grinds the food and is especially strong in grain-eating birds such as turkeys. Many birds with large gizzards have small stones or grit in the gizzard that they have ingested to assist with grinding. It is interesting to note that these birds are able to retain the grit while allowing food to pass through the stomach, thus eliminating the need to constantly search for new grit. In addition to this grit, the inner surface of the gizzard is often hard and keratinized. It contains many folds, providing numerous hard grinding surfaces.

In the gizzard of raptors, fish-eating birds and many insect-eating birds, the indigestible materials (bone, feathers, fish scales, etc.) are compacted into a tight pellet that will later be regurgitated. Owls make the most tightly compacted pellets because most owls swallow their prey whole rather than tear it into smaller bites as many other raptors do. Additionally, an owl's stomach is far less acidic than is the stomach of most diurnal raptors and as a result, bone and fur are not dissolved, but regurgitated nearly intact. Thus, the pellets contain fur, feathers and bones of their prey and teasing them carefully apart will often reveal nearly complete skeletons of the small rodents or birds that the owl fed upon. Hawks and other raptors also form pellets but they are not as well-formed or as complete as those of owls. The pellets of most fish-eating birds are even less discrete, consisting mostly of fish bones and scales. Such pellets disintegrate rather quickly and are seldom found unless one knows exactly what to look for and when or where.

16

In Praise of "Commonness"

If you asked your friends for a list of common birds, chances are the American Robin would be on most lists. Most of us learned about robins in school and we often see them referred to in literature and poetry. But what do we really know of this uniquely American bird, aside from the fact that it has a red breast?

The American Robin is North America's largest, and probably our best-known, thrush. Its brick-red breast and willingness to share human environments make it familiar to even those of us who are not particularly interested in birds. American Robins are found from Alaska and Canada in the north and range south through Mexico. Many of us were taught the story of the "first robin of Spring," which says that when this migratory bird is seen, spring is on its way. Its Latin name is *Turdus migratorius*, which literally means "migratory thrush."

But this little story, taught to so many children,

American Robin, with a worm...or two.

can lead to great confusion. If you live in northern latitudes, such as in Canada, the robin may indeed be absent in winter and be an early spring arrival. But many of us can find robins in farm lots, woodlands and back yards at almost any time of year, so does it really migrate? If you live in parts of the Southwest or in northern Mexico, most of the robins you see year-round don't migrate. But for most of North America, American Robin populations are migratory. Thus, those robins you see looking for worms in your yard in June are not likely to be the same individuals you see in the fields in December. In many parts of the country, robins that spend the breeding season in one location, may spend the winter in another geographical region, and the robins that are seen in the winter often breed somewhere further north. But many of these details are guesswork on our part, because we still know relatively little about the exact movements of robins in many parts of the country.

Now, think for a moment about the robins you have seen. Did they run or hop? Was their beak yellow or dark? And if you have noticed them when they stop on your lawn and cock their head to look down, do you think that they were *listening* or *looking* for worms? Most of us have seen these behaviors, but we may not have thought to ask these questions, nor have we likely watched closely enough to be able to answer them. In fact, robins mostly run when on the ground, usually in short, quick spurts. In adult robins, the beak is mostly yellow with some darkening at the tip. How they find worms may not be so easily answered and you may have heard either answer—by sight or by sound—as the "correct" explanation. Early experiments (and perhaps, common sense) suggested that robins see the worms. But more intricate experiments done in the late 1990s revealed that the answer was more complex. It may not seem like a worm could make any noise when moving through the soil, but the worm's movements do create some very subtle sounds. When the robin cocks its head near the ground, it is actually listening for the worm... but that's not the end of the story. Subtle sounds alert the robin to the presence of a worm, but to be able to capture it, the robin has to finally see it. So the answer to the question is both—it first uses sound, then sight.

The American Robin can also serve as an excellent example of a problem typical of many familiar things—commonness. Like so many things we take for granted, it's easy to recognize a robin and, as a result, we usually don't look at it carefully. And when we don't look carefully, we sometimes don't really know much about what we are looking at. In the case of the robin, we see a large thrush on the lawn with a red breast, so we check it off in our minds as a robin and tend to look no further. Were you aware, for example, that a robin has a white ring around its eye, but a split ring, not a continuous band of color? Look even more closely and you will see white streaks in the throat and a speck of white on each corner of the tail.

Take a few moments to really study a robin the next time you see one, and perhaps you will observe a behavior you have never seen before. Then, extend your observations to anything else that you commonly look at—or that you perhaps look at without paying close attention. If you allow yourself to really see past the commonplace, a new world of exploration and discovery may be yours.

I have always loved to see and share nature and I have been actively birding for nearly 40 years. But only recently did I ever see the courtship dance of the male Lazuli Bunting, a relatively common summer visitor to Oregon. I was with students in my ornithology class, walking along an old roadbed in one of our local county parks. A male Lazuli Bunting sang from the tops of the tall grasses along the edge of the road. I have seen many of these birds, but because he is so gorgeous, and because I wanted to be sure all of my students had the opportunity to see him well, we stood quietly and watched. His singing intensified, and then we saw why. Across the road was a female whose attention he was seeking. Then, the unexpected happened: he leaped to the ground in front of the female, erected his tail, fluttered his wings and danced— swaying back and forth, singing all the while. What an incredible display to witness, and from a very beautiful bird! Because it is a common summer bird that is easily seen, we typically enjoy the male's beauty as he sings from an open perch. But all too often, we pause only briefly, before moving on to the next bird on the path. Had we done so on that day, we would have missed his wonderful performance. What else have we missed in our haste?

There are many things in nature that we consider common or ordinary. We may acknowledge their presence, but we often fail to pause, wait, and look closely, and thus, we fail to fully understand how complex the commonplace can be or what we can learn from it. So the next time you see a robin, stop to study it just a bit longer. See if it behaves differently than you expect or if it can show you something different to appreciate. And when it runs across your yard and stops, looking downward, will you see it snatch a worm from its hole, or devour a tasty beetle instead? There is always so much to learn and it is always fun. Enjoy those robins!

17 Like Water Off a Duck's Back

When it rains, we usually head indoors, but what do birds do in wet weather? Since birds have a high metabolic rate and need to seek plenty of food each day to fuel that metabolism, they can't afford to wait for the rain to abate so that they can continue their normal daily activities.

Two different factors would seem to affect birds during wet weather: their food, especially insects, may be more difficult to find; and the birds themselves would get wet. Let's look first at food availability.

During rain, especially heavy rain, many flying insects seek shelter and are more difficult for birds to find. The peak of our rainy season coincides with colder weather and during these months, many adult forms of flying insects are not present, although their eggs or larvae often over-winter. Flycatchers, many warblers, and other birds that depend on flying insects for food, have usually flown south to warmer environments where flying insects remain abundant. But even during cool, wet weather, there are many insects that do remain active on the surfaces of vegetation, on the ground, and among the detritus on the ground below trees and shrubbery. Bushtits flit from bush to bush seeking the insects that reside on the undersides of evergreen leaves or on the surface of winter twigs. Spotted Towhees and Fox Sparrows scratch at the ground, turning over decaying vegetation in search of insects and other invertebrates or seeds that may be present. Insect and spider eggs are a common food for birds in the winter. Hedges and other dense shrubbery in your yard help provide a haven for insects, spiders or their eggs that are necessary for many of our winter-resident birds.

Many of the songbirds that do remain in our area throughout the winter (and there are many) seek seeds or over-wintering fruit to eat and thus do not depend on a steady source of insects. Chickadees, for example, change their diet, eating large numbers of insects during the summer and mostly seeds in the winter. House Finches and goldfinches do not switch diets seasonally, but continue to seek seeds throughout the year. The seeds at your feeders provide them with an easily accessible source, bringing these welcome visitors into our yards.

For large birds, such as raptors, many of their usual food resources still are active during the winter months. Many rodents remain active in winter so raptors that prey upon them have a steady supply of food. Some hawks, like Sharp-shinned Hawks and Cooper's Hawks, feed on small birds, which are still plentiful in cool, wet weather. Sources of open water provide habitat for herons, ducks and other birds that seek fish or other aquatic food, including aquatic vegetation and the invertebrates that forage on such vegetation. Many shorebirds probe mud flats along the water's edge in search of invertebrates to dine on.

So, finding food is usually not a problem for our winter birds

Note how water beads up on the head and face of this Canada Goose.

even in the steady rain. But, do they get wet? If we stay out in the rain, our clothes get wet and we may begin to feel cool, wet, and somewhat miserable. Our skin may be waterproof but we still are uncomfortable. Of course, we know that ducks love water but how do non-water birds cope with being wet? In fact, under normal circumstances, birds seldom get "wet." That is, their skin is not wetted, because it's protected by a barrier of air, entrapped by their dense plumage of feathers. The fine structure of their feathers ensures that water does not penetrate to the skin.

You may have heard that waterfowl and other birds spread oil from an oil gland over their feathers to waterproof them. This may seem contradictory, especially when you consider that birds soaked in petroleum oil spills become wet and chilled. The truth about birds' oil glands is a bit more complex.

At the base of the tail, on the upper portion of the rump, is the uropygial gland, also known as the preen gland or, sometimes, the oil gland. Secretions from this gland are not simply oils but are a mixture of waxes, fatty acids, lipids (fats) and water. As a bird preens, it rubs these glandular secretions throughout its plumage using the top of its head and its beak. This may provide some waterproofing, but the primary benefit of these "oils" is to keep the feathers supple and in good condition. The secretions stimulate the growth of some beneficial bacteria, but mostly they deter the growth of keratin-degrading bacteria, keratin, and fungi. Preen gland secretions also help discourage feather lice and other parasites. All of these actions keep the feathers in good condition and help the bird maintain the very important fine details of their structure. Despite information in the popular press and even in some textbooks, there is little evidence that preen gland secretions provide much, if any, direct waterproofing of birds' feathers. The ability to repel water comes from the fine physical structure of the feather and only feathers in good condition can repel water. For this reason (and many others), preening is a very important activity for birds and one that consumes much of their time.

Some of you may live near or visit coastal habitats, and if so, perhaps you have seen cormorants with their wings outstretched, drying their feathers in the sun. This observation has led some people to an erroneous conclusion, one often repeated in the popular press. It is often assumed, incorrectly, that because the wings get wet, that cormorants do not have a preen gland. In fact, cormorants do have a preen gland, and, although simple in structure, it functions quite well and serves the same purpose as it does in other birds. Most of a cormorant's feathers are waterproof. Without this protection for the body feathers, the bird would lose the insulation necessary to protect it from the cold temperatures of ocean waters in northern latitudes. Only the cormorant's wing feathers are "wettable;" the rest of the body remains dry. The air trapped beneath this dense mat of feathers, which provides insulation and protection, also provides buoyancy. This is a benefit that helps most water birds to stay afloat, but cormorants often dive deeply to feed and any added buoyancy can hinder their deep diving below the surface. The structure of the primary and secondary wing feathers is such that water *can* penetrate, wetting the wing and making it a bit easier to dive beneath the surface of the water. But *only* the wing is wetted and the body remains protected. Thus, cormorants can be seen standing on a rock or other perch with wings out-stretched in the sun to help them dry. But the birds are not disabled during this drying, because even with "wet" wings, they are still able to fly.

The Double-crested routinely performs this wing-spreading posture, and this behavior is much less frequently seen in other species of cormorants. This is also the only species of cormorant found inland, which means that you might very well see this display along rivers, lakes or ponds, especially during the winter when these birds can be quite numerous inland. In some communities, there are known roosting places where cormorants congregate at night. If such a place is nearby, you may be able to watch them as they gather early in the morning or again at dusk.

When you next find yourself out in the rain beneath your

Keratin

Keratin is a complex of proteins produced by skin cells of all vertebrates, and it forms microscopic fibers that combine to give strong mechanical properties to structures such as fur, hair, nails, claws, scales, feathers and bird beaks. There are both alpha-keratins and beta-keratins, but beta-keratins are found only in birds and reptiles. One type of beta-keratin is responsible for the structure of feathers. The differences between all of these keratin types has to do with the particular amino acid sequence of the molecule.

Keratins cannot be digested in the digestive tracts of animals, but some bacteria and fungi in the environment are known to break down keratin. We know very little about which specific bacteria or fungi can do this, but imagine how deep we might be piled in bird feathers if nothing could eventually break them down. Some of the bacteria or fungi that is found in birds' feathers begins to wear away or weaken the feathers if allowed to grow. It is believed that one function of secretions in the preen gland helps keep these organisms from growing on feathers.

umbrella or bundled in your high-tech rain-gear, look for the birds out in the same wet weather. It is then easy to admire how well the natural processes of nature have adapted them to survive wet winter climates.

A Double-crested Cormorant, with its wings spread out to aid drying.

18

Mallard: A Familiar Duck

When we think of a duck, the image that comes to mind for most of us is the Mallard. This is the most widespread and common of all dabbling ducks in the northern hemisphere, and can be found any place a small amount of open water exists. There is evidence for domestication of Mallards in Europe as far back as the twelfth century and in Southeast Asia as long as 2,000 years ago. Most of the well-known and popular breeds of domestic ducks were all derived from Mallard stock.

The male Mallard is one of the most easily-recognized of all ducks, with his bright green head, brown chest, and curly, black tail feathers. The familiar loud quacking is a sound we often associate with ducks, but many people are surprised to learn that it is only given by the female Mallard. The male has a very soft quacking sound as well as a variety of other vocalizations, but not a loud quack, and many ducks whistle or grunt instead of quacking.

A female Mallard sleeps, with one eye seeming to be open and awake.

Like many ducks, a Mallard's daily activity cycle consists of eating, bathing, preening, and sleeping. Mallards are generalists in the foods they seek. During the breeding season, the primary food consists of insects, especially aquatic insect larvae, but they will sometimes snatch insects flying past them, or take insects from the surfaces of plants. They will also eat some snails and other aquatic invertebrates, and earthworms when they can be obtained. During the non-breeding season, the diet is mostly aquatic plants, seeds and grains, including cereal crops, and wheat.

Preening is essential for all birds to keep their plumage in good condition. In aquatic birds such as Mallards, frequent preening is necessary to help maintain feathers' water-proofing. The fine micro-structure of the feathers is what prevents water from penetrating through to the skin, not "oil" from the preen gland. (For added details, see Chapter 20, page 135.)

Ducks spend significant amounts of time sleeping, but sleeping during the day can be risky, especially when predators may

A Mallard drake (male), is one of the most familiar birds across the northern hemisphere.

be nearby. A duck sleeping on the surface of the water, or on the nearby shoreline, must have a way to be alert to danger. Ducks and many other birds have the ability to allow half of their brain to sleep while the other half remains alert. You may notice this if you watch a duck sitting with one eye open and one eye shut. The open eye and active portion of the brain remains alert, and if danger is seen, the whole bird awakens, ready to flee if need be. This is not simply a matter of closing one eye, as appearances suggest. Experimental measurements of brain activity reveal that half of the brain is truly asleep.

From winter through spring, the normal daily routine is interrupted by courtship and breeding activities. Spring is the time for nesting and egg-laying for Mallards, but breeding activities begin long before then. Although Mallards are monogamous for a season, they must establish new mates each year. Courtship and pair formation can begin as early as September, and during the late fall and winter months, male ducks of most species are in their finest plumage and compete for the attention of females. Apparently, female Mallards are most attracted to males with the brightest

Male Mallard (top left) performing the grunt-whistle display as part of the courtship ritual.

green heads and brightest yellow-green beaks. As many as five males sometimes surround a female, all vying for her attention. They perform a ritualized display, raising the head high into the air, then bobbing down with tail raised higher, all accompanied by a series of whistles and grunts. These behaviors are so stylized and so identical between males that biologists refer to specific portions of this dance as: Grunt-Whistle, Head-Up-Tail-Up, and Down-Up. The female makes the ultimate decision about a mate when she swims away with the male of her choice. Because the actual mating appears to be rough and forceful on the part of the male, it is termed a "rape" in many duck species. Early in the season, pairs may not stay together, but eventually a strong bond is formed between the two birds and when spring arrives, the male simply follows his chosen female to her breeding site. The female selects the nest site and builds the nest. At this point, the pair-bond begins to break down. The male may stay in the area but plays no role in caring for young. The "families" of Mallards that you may see together are not really families, as the male may or may not be the female's mate. In either case, the male has no concern for the young, but is simply seeking another mating opportunity.

In rural and more natural settings, Mallards seldom re-nest in a single season. In urban environments, however, Mallards may re-nest and the more domesticated they become, the more their behaviors differ from purely wild birds. Unfortunately, many females make poor choices for nest location, sometimes selecting a back lot far from water, making it necessary to lead the newly-hatched chicks across busy streets to reach the water's safety. Some have even selected flower-boxes on window ledges several stories above the ground. More of these poor choices are likely as the human and Mallard environments continue to overlap.

Whether or not you are deeply interested in birds, you most likely are familiar with the Mallard. It is so common that we tend not to look closely at all of its features. But do take a moment to watch them at your next opportunity and you will see how beautiful and interesting they really are.

Keeping Cool

When the summer weather is upon us, daytime temperatures often reach into the 80s or 90s in many parts of the country. At such temperatures, we certainly would not pull a down parka out of the closet to wear outside. Yet birds must wear their coats of down and other feathers every day of the year. With all of this insulation, how do birds manage to stay cool in hot weather?

It may seem to you that it would be a good idea for birds to simply have fewer feathers during summer, and in fact, that is the case for some birds. But such an adaptation is mostly found in birds that live in environments with very cold winters and warm summers, birds such as ptarmigans, a grouse that lives at high elevation. Most familiar yard birds wear approximately the same number of feathers throughout the year. Interestingly, even in hot weather, small birds such as chickadees spend more energy keeping warm than cooling off. This may sound strange, but smaller birds have more surface area in proportion to their size than do larger birds, so they lose heat more quickly and therefore need to find ways to keep warm.

Birds have faster metabolisms and maintain higher body temperatures than humans, so if we had the same body temperature as is normal for birds, we would be running a dangerously high fever. A song bird's average body temperature is about 105°F, so while outside temperatures of 95°F seem hot to us, it is well below the body temperature of a bird. Despite this, birds do get hot and must cool off.

As air temperature climbs, most of us begin to perspire. Even tiny drops of sweat evaporate as they reach the surface of the skin and this evaporation has a cooling effect. But birds don't have sweat glands so they must find other ways to cool off using evaporative cooling. The most common method of doing this is panting, just like a dog does. Panting moves air across a moist surface (the mouth and respiratory system) and has the same cooling effect as sweating. But panting has some disadvantages, which birds must compensate for. The more a bird pants, the more body water it loses. There are only three ways to replace water lost in this manner: drinking, obtaining water from food, or getting water as a by-product of natural metabolic processes. Doves and pigeons, for example, need to drink water every day and in some dry environments, doves fly many miles each day seeking water to drink. Goldfinches can extract some water metabolically even from seemingly dry seeds.

Water features can be very attractive to birds and many different kinds of birds will come to drink or bathe at them. Although birds can obtain moisture from foods they eat, especially from insects, drinking water is easier and more efficient, when available. In addition to the chickadees and finches that routinely come to feeders, birds you might not otherwise see will be attracted to water if you provide it. Several kinds of sparrows and warblers, for example, may

Mourning Doves in dry environments fly daily to water sources.

come to a water feature but not be attracted to feeders. A shallow dish or bird bath is best, or a pool with shallow water near its edge. Very shallow water (~1 in.) will allow birds to wade, wetting the breast feathers. The process of allowing water to dry from birds' dampened feathers helps cool them, and drinking helps replace water lost due to panting. The sound of moving water is often more attractive to birds than static pools. Drippers are popular and can be purchased to attach to most bird baths. They provide a steady drip of water, the sound of which alerts birds that water is available. More elaborate features can have water spilling over rocks or into a pool sounding like a natural stream. Be sure that the water you provide is clean and fresh. Particularly in bird baths, water can become stagnant, and quickly fouled when used by multiple birds.

Fat, such as is found in suet, is a very high-energy food, but a bird's body can also use fat to produce water during metabolism. For every gram of fat a bird eats, it will produce just over a gram of water internally, and this can be important at times when water is difficult to obtain, such as during dry summer months. In contrast, a gram

A Chestnut-backed Chickadee sunbathes on the railing of our deck. Such activity may help the bird get warm wjuickly and the heat can cause feather parasites to dislodge, making it easier for the bird to remove them.

of carbohydrate will produce only about one-half gram of water and even less water is obtainable from proteins. So, offering suet to birds can also provide them some water. However, be cautious about providing suet during really hot weather. Make sure your suet is fresh and has not become rancid, and avoid letting the heat melt it so it drips. Soft suet can become matted in feathers, especially around the face. Woodpeckers have been reported to have lost many feathers in the nostril and around the face and beak as a result of becoming coated with soft suet.

Shade can often provide enough temperature difference to help birds stay cool. Nestlings just a few days old are not able to fully regulate their body temperature, so the adult female (and in some cases, the male as well) broods, or sits, over the youngsters to help keep them warm. But as temperatures climb, the adult's heat can become too great so they will raise themselves above the young to provide only shade. Sometimes adults spread their wings to provide additional shade for the young, but wing-spreading also helps adults keep cool by providing more surface for air to circulate through. Birds will seek shelter in dense bushes, trees or other vegetation, where the cover of the leaves keeps the air temperature a few degrees cooler than in the sun, and that difference may just be enough to keep the bird cool. Bushes also provide the additional benefit of safety if the bird is not visible to potential predators, and it may sit quietly, avoiding unnecessary energy expenditures. Thus, your trees and shrubs could be assisting birds that aren't at your feeders.

As strange as it sounds, black feathers can also help birds stay cool. The dark melanin pigment in the feathers absorbs heat, but the feathers are not pressed tightly against the skin, so the feathers become warmer than the skin below and this small temperature gradient creates a slight movement of air beneath the feathers that draws heat away from the skin and helps cool the bird. Turkey Vultures, Ravens and other birds of desert environments are typically black or dark over much of the body. Conversely, Arctic birds are often white, allowing the sunlight to penetrate and directly warm the skin, which is often black in these birds.

When you see a Turkey Vulture with its wings spread, it is usually warming itself. And, there is another advantage to spreading the wings. If the feathers can be sufficiently warmed, it can be irritating to parasites within the plumage. They may loosen their hold and move to the shady side of the feathers. While they are loose, the bird can more easily remove these parasites when preening. You may be able to observe this in many different species. I have seen this behavior first hand in jays, Brown Creepers, chickadees, and a few others and it has been reported in many species worldwide.

It is important for birds to maintain consistent temperatures to stay healthy. Providing shelter, shade and water are some simple ways you can help them and these have the added benefit of allowing you the opportunity to see and enjoy some of these colorful and fascinating wild neighbors.

20

Have You Seen a Weary Willie?

So, you've gathered together a marvelous collection of old bird books. It's fun to read, but I'm not sure I have ever seen some of the birds named in them. Maybe we should try to find them.

It's a pleasant spring day so let's do some birding at a nearby lake and marsh. Look—there's a *Lightwood Knot* swimming by, thumping its breast with its bill in search of a mate. In the cattails above it, I see a male *Field Officer* displaying and calling to defend his territory, and in the distance I can hear the strange booming of a *Punkatunk*.

You say that you can't find any of these birds in your field guide? Try *Shot-pouch, Maize-thief,* and *Look-up*. I guess that's not much more helpful. Try again but this time search for Ruddy Duck, Red-winged Blackbird, and American Bittern. That's better. Now we all understand what we are talking about. The problem was those pesky

The Steller's Jay is sometimes called a "Camp Robber" by human campers.

local names. They may be very colorful, but not everyone knows what we are talking about when we use them.

Unlike many things in nature such as wildflowers and trees, birds have a set of official, consistent English names that nearly everyone uses, at least in print. These names, found in the *American Ornithologists Union (A.O.U.) Checklist of North American Birds*, are used in field guides and most current publications about birds. However, the AOU reviews the literature and revises some of the taxonomy each year. (A supplement is published in July or August.) So some of the names used in your field guide may have changed, even if your guide is only 2–5 years old. But old, and long-used names tend to stay with people, and regional and colloquial names are still frequently in use, especially by hunters and "non-birders."

We all believe that we know what bird we are talking about

when we refer to a robin (or more properly, American Robin), but what springs to mind when I refer to a blue jay? To birders, a "Blue Jay" is the Blue Jay (*Cyanocitta cristata*) of eastern North America. However, many people enjoy feeding birds but don't necessarily know details, or even the real names of many of the birds visiting their yards. In the West people may refer to the Western Scrub-Jay (*Aphelocoma californica*) or the Steller's Jay (*Cyanocitta stelleri*) as "Blue Jay." All three jays are blue but only one of them is officially referred to as Blue Jay.

Standardization of names is useful and helps clarify our discussions. Without this standardization, even "robin" could be a problem. To anyone in North America "robin" invariably means American Robin. But what if you are talking to someone from Europe? When you say robin, unless they have been in this country a long time, their mind will most likely bring up an image of the European Robin, a very different bird. Elsewhere in the world, you will find other "robins," for it seems that everywhere the British went in their early explorations, they gave names to birds that reminded them of their own homeland birds. The European Robin has an orange red breast. So when they saw any similar bird with red on the breast in foreign lands, it was called "robin." Thus, we have a robin in North America but there are also robins, and more than one species, in India and Australia as well. Robin means something different depending on what part of the world you are in or from.

When looking at the names of North American birds, we find that many of the old names were, and in some cases still are, in use in only a small region of the country. Other names had much broader usage and were once part of the standardized list of names. For example, the fourth edition of the AOU Checklist (1931) lists Holboell's Grebe and European Cormorant. Today, we call these same birds Red-necked Grebe and Great Cormorant.

You will also find such names as Baird's Wren, Nicasio Wren

and Vigors's Wren, all of which are subspecies of Bewick's Wren. Recent editions of the Checklist no longer assign common names to the many different subspecies. Some names have changed so completely that you may be at a loss to find them without great difficulty. For example, your field guide most likely does not list a Cactus Woodpecker, *Dryobates scalaris*. That's because both the English and Latin names have been changed. Your field guide should now call the Cactus Woodpecker a Ladder-backed Woodpecker and *Picoides scalaris* will be the correct Latin name.

Many names are simply a good general description of a bird, or at least describe one aspect of how it looks. Yellow-headed Blackbird, for example, very aptly describes the male of this species, which indeed does have a bright yellow head. Black-throated Blue Warbler is a good visual description for the male of this eastern warbler, which has deep bluish upper parts and a black throat and sides. However, this description only applies to the male. (See Florence Merriam's quote about this in the essay *What's In a Name? Chapter 25*).

Other names are imitative of the sounds that birds make. It would be an absurdity to think that a small plover would prey upon deer as the name Killdeer might suggest. Killdeer is a "sonic" name—the word is a common call of this plover. Sora, kittiwake, chickadee and cuckoo are also names describing vocalizations.

I began this essay by referring to the Ruddy Duck as a *Lightwood Knot* and a *Shot Pouch*. How do these names describe this small, diving duck? Where do the names *Field Officer* and *Maize-thief* for Red-winged Blackbird come from and why is the American Bittern referred to as *Punkatunk* and *Look-up*? Let's take these, and some other unusual bird names, and examine them more closely.

The Ruddy Duck is a small duck belonging to a group of ducks often referred to as "stiff-tailed" ducks. Ruddy is a

"Field Officer" or "Maize-thief," are names given to the Red-winged Blackbird, a male shown here.

good description: this rusty color is prominent in the summer plumage of the male. Males in summer also have a bright blue bill and Bluebill is another old term sometimes used. But Bluebill is a confusing name, as many hunters more commonly refer to scaup and Ring-necked Ducks as *bluebills*. Over the years, hunters have come up with many colorful terms for ducks, but the Ruddy Duck has more colloquial names (I have found over 80) than most other ducks. In old references, the claim is made that this duck is hard to kill, which has given rise to names such as *Shot Pouch*. It was claimed that a Ruddy Duck can absorb a whole pouch of shot without being killed. A hard-to-kill reputation also gives it the name of *Iron-head*, *Tough-head*, and *Hickory-head* (hickory is a very hard wood). *Lightwood Knot* has the same meaning. Lightwood is a term used to refer to the resinous pine wood from which turpentine is derived, and the knots in this wood are said to be especially hard, thus the name *Lightwood Knot*.

Dickey (also spelled *Dicky* and, occasionally, *Dinky*) is

A Shot Pouch, Iron-head or Hickory-head... also known as a Ruddy Duck.

another delightful name sometimes given to the Ruddy Duck. To explain it, I can do no better than to quote from *Names and Portraits of Birds* by Gurdon Trumbull, 1888:

> *... and one Wilmington ducker told me of hearing the Ruddy called Dickey by certain South Carolina gunners—"Don't you know," said he, "how, when they start, they go dickey-dickey-dickey, patting the water with their wings and feet?"*

As in the above example, it is often an action that gives a bird its name. It is obvious to most of us how a woodpecker acquired that name, and sapsuckers really do lick sap from holes that they have drilled into the trees. Behavior led some people to call the Red-winged Blackbird *Maize-thief*. In late summer and early fall, blackbirds begin to gather in large flocks, and these may wander through agricultural fields, some of which are corn or maize fields. While they may eat some of the grain (especially the dried grains remaining after harvest), most of their foraging in these fields is probably for insects. It must have appeared that the blackbirds were stealing the corn from the farmers so they were called *Maize-thief*. When you see these birds flying into the fields, you can't help but notice the bright red patches on the wings of the males and it is these feathers that give the bird the name used by birders and ornithologists today—Red-winged Blackbird. But, these red patches are also suggestive of the epaulets on the shoulders of some military uniforms, and that's how the old name *Field Officer* came into use.

The American Bittern has a behavior that has earned it the name *Look-up*. This heron of the marsh conceals itself amongst the bulrushes and cattails, and its eyes are located rather low on the side of its head, so when its beak is pointed straight forward, its best field of view is downward, enabling it to easily see the marsh floor as it forages. But to see approaching danger more clearly, the bittern may point its beak skyward so that the downward pointing eyes are now directed forward to see on-coming danger. As a result of this pose, it has been dubbed *Look-up* and *Sun-gazer*.

For many people, the sound of the American Bittern's call is one of its most intriguing and endearing features. This strange, pumping call is the source of many local names, some of which are sonic in origin, such as the aforementioned *Punkatunk, Oomptah, Bill-gudgeon, Plum-gudgeon,* and *Dunk-a-doo*, each of which attempts to mimic the bittern's call. To some people, the call has suggested an action, which is reflected in names like *Barrel-maker, Devil's Pump, Pump Bird, Pile-driver,* and *Thunder-pumper*.

Like other herons, the bittern makes quick stabs into the water with its beak when foraging. This has earned it the name *Poke*—a name often applied to other herons as well. The Great Blue Heron is sometimes called *Poke, Big Poke, or Frog-stabber*, and the Green Heron is also called *Poke* or more frequently, *Shite-poke* (the polite form of *Shit-poke*). The latter name is applied because when a Green Heron flies from danger, it often defecates, leaving a trail of white, explaining the origin of another of its names, *Chalk-line*.

Sometimes, a whole group of birds will have an odd name. The Family Caprimulgidae includes birds such as nighthawks and Whip-poor-will (another sonic name). Birds in this family are commonly referred to as "Goatsuckers," which seems like a very strange name for a group of birds whose diet consists of insects caught while on the wing. But, the name has been handed down from old European folklore, probably the result of poor observation and a misunderstanding of the habits of these birds. These birds have tiny beaks but very large gapes, revealing large mouths when open, perfectly adapted for catching insects in flight. The birds feed while in the air but many nest on the ground. If startled while at rest upon the ground, and if unable to immediately take flight, they will try to defend themselves with their mouths wide open. Somewhere in the past, this action must have led someone to the erroneous conclusion that these birds leap up from the ground and attach themselves with their mouths to the teats of goats and cattle to suckle the milk. It is an odd behavior to ascribe to any bird and in fact, does not exist; yet the name persists.

There are many delightful names that can be found for small birds as well. The Song Sparrow has a marvelous song and thus the bird has sometimes been called *Silver Tongue* or *Everybody's Darling*. The Wood Thrush has a glorious, rich musical song and has been called *Flute* and *Hautbois* (a hautbois is an oboe). The opposite is true of the Cedar Waxwing which has a very weak, thin voice, totally inappropriate for its sometime nickname, *Carolina Chatterer*. The *Polite Bird* is a name that better suits this bird because waxwings sometimes pass berries or small fruits back and forth between individuals perched together. This must have seemed a very polite behavior to whoever first used the term. A *Devil-down-head* is a nuthatch (usually White-breasted or Red-breasted) and is a reference to the fact that nuthatches can walk head-first down the trunk of a tree. Nuthatch, itself, must seem like a strange name, suggesting that these birds try to hatch nuts, which would

The Common Nighthawk is a member of the "Goatsucker" Family.

be a complete absurdity. The name is actually a corruption of the name used in Middle English, which means nut-hatchet. Nuthatches will sometimes take a small nut or seed to a branch and hammer at it to open it. So, nuthatch does suggest a true behavior, just not hatching nuts.

The origin of some names is lost and we are left to ponder how these came to be. Why is a Cedar Waxwing called *Blockhead*? Did its crest suggest a blocky appearance to someone? Why should a Loggerhead Shrike (a predatory songbird) be called a *Cotton-picker*? And why should a Northern Shoveler have been called *Cow-frog* or *Soup Lips*? Could you easily guess what bird is called *Pork-and-beans*? It certainly is a strange name for a Common Nighthawk.

There are thousands of other names that have been used, but for now, I'll just give you one more. My title asks if you have seen a *Weary Willie*, but how many of you know what a *Weary Willie* really is? Those of you who live along the west coast of North America are probably acquainted with this bird, at least during the winter. You most likely know it as a Golden-crowned Sparrow. (For more information about these birds, turn to the essay on Golden-crowned Sparrows, Chapter 10, A Western Specialty) The song of the Golden-crowned Sparrow is a plaintive, slow descending series of whistled notes, sometimes said to be saying "oh-dear-me." It does have the quality of someone who is perhaps just a bit tired, hence the name *Weary Willie*. (We humans do like to be alliterative in our names—somehow *Weary Frank* just doesn't have the same ring to it!)

Birding has become very popular and is one of the fastest growing recreational activities in America today, with 68.6 million Americans watching wildlife around their homes. Of those people, over 52.8 million feed birds and wildlife (2011 US FIsh & Wildlife National survey). Finding old and local names and examining their origins lends another element of fun and discovery to that pastime.

Now, the day is still young and I can still hear the *Punkatunk* booming out in the marsh. The *Field Officer* still proclaims that this patch of cattails belongs to him. The *Lightwood Knot* is out of sight but something is coming out of the reeds over there. Can you see it yet? Yes, I think you are right. It does look like a *Frog-in-throat* (Pied-billed Grebe)

21

Biobigotry

Starlings at the suet feeder! Cowbirds in the neighbor's trees! What is your response? Do you utter a mild curse under your breath while you step outside to chase them away? Perhaps you are dreaming up new ways to keep them at bay while you think hostile thoughts about them and their kin.

We tend to instantly dislike birds such as cowbirds and starlings, labeling them as "bad," while giving them no respect. Similar sentiments are expressed all across the country every day by birders and backyard bird-watchers. Then there are the "dirty pigeons" and "lazy House Sparrows" that somebody should do something about. A few people go a step farther and justify to themselves a need to shoot hawks and eagles. But all of these statements are simply judgments without thought, based on little factual information. Some people have called this "biobigotry," a term that I find very apt and one that can have

The Bald Eagle will dive on an Osprey carrying a fish in an attempt to steal the fish for itself.

the effect of making people stop to think. Why do we so dislike starlings, cowbirds and some other species? Are we somehow privy to a higher moral code that all species must follow, lest they be considered villains?

In the deserts of the Southwest, there is a wasp that hunts tarantulas to paralyze them with a sting. It then lays its egg on the helpless spider which then becomes an immobile but living food source for the wasp maggot. This may offend our sense of ethics but should we label the wasp as a cruel and evil creature that has stepped beyond the bounds of decency? Some Victorian writers did just that, but I would hope that today, we can look beyond any personal revulsion to see this as one of the many ways that evolution has shaped the natural world. It is amoral: neither evil nor good, but is simply an efficient way to compete for the right to continue your species.

If we are willing to examine life closely, we will find negative values in the "good" species and positive values in the "bad" species. We are thrilled to see a Bald Eagle, but it is a thief, stealing food from others whenever possible. Thievery in eagles is simply a good survival strategy, not a criminal way of life. Many small, "cute" rodents in some areas, including near us in eastern Oregon, carry bubonic plague, so are they really evil? This rarely affects humans since their fleas (which are needed to transmit the disease) very rarely bite humans even when human contact occurs. (Recent incidences of plague occurred when a man's cat was in contact with rodents; another really good reason to keep cats indoors.)

Western Bluebirds are one of the primary vectors for the spread of mistletoe in oak trees. We know mistletoe is a parasite (is it therefore evil?) but we love bluebirds. And who is not excited to see, at a distance, a bear dining on huckleberries? But in reality, the bear is destroying some of the future progeny of that huckleberry plant—its seeds. Does this make the bear evil or as offensive as the wasp who destroys a tarantula? Neither the eagle, bluebird, bear or wasp should be judged by our artificial moral code; they are simply doing what nature has best equipped them to do. The same is true for starlings, cowbirds and other living creatures that may do things that offend us.

Starlings and cowbirds are both very skilled at survival and both have fascinating adaptations for success. They are all too often portrayed as evil, harmful creatures but much of their current success is due to our manipulation of the environment.

Starlings are much-loved in their native Europe and they can actually be very beneficial because they often eat many of the insects and insect larvae that damage crops. Starlings also seek out cranefly larvae, a common lawn pest, and can quickly destroy large quantities of them, as well as insects that are considered to be agricultural pests. Yet this is never talked about when people discuss starlings. They are viewed in such negative terms that few people stop to examine the complexities of their existence or allow that they may have some beneficial effects on the environment. Nothing in nature is purely good or evil. But, even otherwise competent people sometimes fall into this trap. Just look at the words of Frank Chapman, often considered the founder of modern

American Ornithology. In his book *Birds of Eastern North America*, (1902) he says the following about cowbirds:

The Cowbird is an acknowledged villain, and has no standing in the bird world. English Sparrows, either because they are not aware of the customs of New-World bird life, or because of a possible and not unlikely affinity, associate with him; but no self-respecting American bird will be found in his company.

As an outcast he makes the best of things, and gathers about him a band of kindred spirits who know no law. There is an air about the group which at once tells the critical observer that their deeds are evil. No joyous song swells the throat of the male. His chief contribution to the chorus of springtime is a guttural bubbling produced with apparently nauseous effort...They build no nest, and the females, lacking every moral and maternal instinct, leave their companions only long enough to deposit their eggs in the nests of other and smaller birds. I can imagine no sight more strongly suggestive of a thoroughly despicable nature than a female Cowbird sneaking through the trees or bushes in search of a victim upon whom to shift the duties of motherhood.

How could such a respected scientist and observer have such a biased view of the natural world?! That was more than 100 years ago, yet I still hear birders today who espouse the same basic attitudes. This, despite several studies showing that cowbirds typically have little effect on populations of most native birds. Only in isolated instances, such as with Kirtland's Warblers, do cowbirds have

Because European Starlings are a non-native species that competes for nest cavities with native birds, they are disliked by North American birders.

a major impact. Host species that are strongly impacted by Brown-headed Cowbirds are ones that are already threatened for other reasons (habitat loss, etc.) and the cowbird is just one more source of pressure, the proverbial "straw that broke the camel's back."

If we are willing to stop and examine nature in all of its complexity, we will discover many fascinating things that we may have never seen before. We will find that each creature is beautifully adapted to live as it does. The activities that we engage in are often at fault for altering the delicate balance that keeps an ecosystem healthy and properly functioning. Starlings would never have succeeded in this country if we had not cut the woodlands and erected structures that gave them nesting success. What we found beneficial matched the needs of the starling. We are to blame for its spread; the starling is only using the tools nature equipped it with.

The next time you see a fly, don't swat it. Instead, watch it and marvel at how it can make twists and turns in the air, often avoiding being hit by your hand. Then watch as it comes to a stop, alighting upside-down on the ceiling. How did it do that? A barrel-roll or a loop-de-loop?

Is a fly "Evil?" It can do things that we cannot, like fly or walk along a ceiling, up-side down.

It actually does neither! As it approaches the ceiling, flying right-side up, it lifts its two front legs up above its head. When these legs make contact with the ceiling, they engage the ceiling's surface and cause the fly's body to flip up onto the ceiling.

Even more wondrous is the fact that it stays on the ceiling without falling off! I am quite certain that is something none of us can do.

Let no creature fall victim to our misplaced biobigotry, be it a fly or an eagle. Instead, study it and find something truly intriguing about it without looking through the filter of an artificial moral code that finds some creatures "good" and others "evil." You might still swat the mosquito biting your arm, but perhaps you can appreciate that it, too, has a place and is only doing what is best to assure the survival of her offspring.

22

What's Your Favorite Bird?

Whenever I give public lectures, lead field trips or teach classes about birds and natural history, I am asked many questions. One of the most frequently asked is, "What is your favorite bird?" How can I answer such a question? Is it the most beautiful, the most impressive, the one with the best song? There are many candidates.

Perhaps I should choose the Raven—a most impressive bird. It is the largest songbird in the world but it has many qualities besides size that make it a good selection for "favorite." Ravens can be long-lived, living perhaps up to 40 years (at least in captivity) and they are adapted to a wide range of habitats. They can soar high above the forest, eat a wide variety of foods and they have few predators. Ravens are also among the most intelligent of all birds, but these are just a few qualities that could qualify them as a "favorite" species. But, once I select this as my favorite, other birds come to mind.

Bushtits are delightful to watch as they cluster around a suet feeder.

How about Bushtits? What could be more wonderful than looking out my window as four or five Bushtits emerge from the brush and cling to the suet feeder? But, a moment later, the whole suet feeder is suddenly covered with Bushtits, their slender tails pointing in all directions. Are there 10, 20? Before long, there are 40–50—or more. How do you count such small birds in constant motion, first covering the feeder, then suddenly dashing into nearby shrubs in a bobbing, weaving mass of small gray feather balls. In the breeding season, these flocks break up and you find them mostly in family units. Could anything be more fun than watching young Bushtits emerge from their nest, that incredible structure, like a woolen slipper hanging neatly from a small branch? I will always delight in seeing a group of young Bushtits huddled together, lined up side-by-side on a branch. Each sits with their mouth agape while the parent patiently goes down the line, filling each mouth. This is

Snowy Egret: Another of my favorite birds is this beautiful, graceful bird

surely a worthy choice for "favorite bird."

But let's not forget the Lazuli Bunting, that magnificent turquoise jewel sitting atop a bush and singing as though song was his greatest joy. It is such fun to hear exclamations of delight when you have shown someone this bird for the first time: "I had no idea that such a beautiful bird lived here." "WOW! He's gorgeous!" A Lazuli Bunting always is a wonderful sign that spring has finally arrived.

If Spring has arrived, then there must be another favorite nearby—listen a moment. Do you hear that rich, melodious up-and-down song? Spring is definitely here, because that's the Black-headed Grosbeak singing its up-lifting and heart-warming song. And when you see the male, he's just as gorgeous as his song! The female may not be quite as bright but she is very attractive in her own right and if you watch her long enough, she just might reward you with a song that nearly equals that of her mate's.

That's right. I said a female singing. Maybe you have heard or read that only male birds sing. As a general rule, that is true but there are a few species in which females don't simply

Unusual among songbirds, the female Black-headed Grosbeak has a song much like that of the male.

have calls but sing like males, and Black-headed and Rose-breasted Grosbeaks (found in eastern North America) are two such species. I'm sure this must be someone's "favorite bird."

How could I not consider the Red-breasted Nuthatch? Here is a bird that can not only climb up the trunk of a tree but it can turn up-side-down and walk head-first down the trunk of a tree. How special is that? Sometimes, you may hear it high in a tree continually repeating its nasal "ank, ank, ank" call, alerting everyone that some danger is nearby. Perhaps it is warning that a Northern Pygmy-Owl is nearby, a tiny, day-hunting owl that feeds on small birds. If so, watch when the owl flies and you may see it followed by the nuthatch and perhaps a few chickadees or other birds, harassing the owl as they try to drive the danger away. Of course this line of thought could easily lead me to name the Northern Pygmy-Owl as a "favorite bird."

There are so many choices, that I could go on and on selecting "favorite birds." So what do I answer when asked, "What is your favorite bird?" The best answer that I can think of is, "My favorite bird is whichever one I am watching—or listening to—at this moment."

Maybe you are like me and have a difficult time selecting a favorite bird. But I think it is okay to have many favorite birds and perhaps your favorite just now is the one feeding at your bird feeder. As a favorite bird, look at it closely. You might see something about it or see a behavior that you have never seen before. If so, you will now have an even bigger reward for selecting this "favorite bird."

23

Looking for Yellowheads

It was mid-July in 1973 when a graduate student in our department came to ask if I would like to help him on a photography trip he had in mind. Birds were not his field of study, but he was very interested in birding as a recreational activity and he knew that I also had a keen interest in birds. Jay had just heard about a local colony of Yellow-headed Blackbirds and he needed a canoe partner to help him go into the marsh to photograph them.

Even today, Yellow-headed Blackbirds are not common in Oregon west of the Cascade Mountains and, in 1973, the breeding colony at Fern Ridge Reservoir west of Eugene was the only one I knew of in the western portion of the state. I have always enjoyed Yellow-headed Blackbirds as they are brightly colored and make what I consider to be delightful "squawks" and "cackles." An opportunity to observe these birds up close sounded like a pleasant way to spend the day and I readily agreed to go with him.

Fern Ridge Reservoir is located about eleven miles west of Eugene, Oregon. It is a shallow, man-made lake and when filled to capacity in the summer, is very popular for sailing and other types of boating. An extensive portion of the lake's southern perimeter is composed of marsh and wetlands and this habitat makes for one of the better birding locales in western Oregon. These marshes are where we were headed for Jay to make his attempts at photographing Yellow-headed Blackbirds. We loaded his canoe on the top of his car and headed out for what we hoped would be a pleasant and successful morning.

In about twenty minutes, we reached our destination, readied our gear and prepared to launch the canoe. Jay decided that the only lens he needed to bring was his 400mm telephoto. He reasoned that since birds were small and always somewhat distant, any other shorter focal length lens would just be unnecessary weight. He also brought only two rolls of film. As I recall, I suggested that he might want to reconsider, as there was plenty of room in the canoe

Extensive cattails or other emergent vegetation may attract Yellow-headed Blackbirds within their range.

for more, but he was sure that this was all that would be necessary. In the pre-digital age, I could not have imagined going on a photo expedition with only two rolls of film, and now I always shoot enough photos to equal many rolls of film (as my wife will certainly attest), but this was all Jay felt he needed and this was his trip.

We put in where Coyote Creek entered the reservoir and followed the stream course as it meandered through extensive cattail marsh before entering open water. I took the rear seat so that I could position and steady the canoe to give Jay unobstructed opportunities for getting his photos. Barn Swallows and Tree Swallows filled the air overhead while Red-winged Blackbirds and Common Yellowthroats called from the surrounding vegetation. As we continued floating lazily into the marsh, Marsh Wrens often hopped up to offer their opinion of our intrusion into their territory. I love watching Marsh Wrens: the male always seems so alert with his tail cocked high over his head giving off a rapid chatter that seems to involve his entire body. However, Marsh Wrens don't typically stay still very long and while Jay was snapping lots of pictures, I suspected that many of them were of blurry spots in the vegetation. The birds were moving, Jay was hand-holding his long lens in a moving canoe and the birds were still somewhat distant even for a telephoto lens. Jay's enthusiasm sometimes exceeded his skills as a photographer, but I gave him high marks for enthusiasm. Finally, one wren sat still in clear view for Jay to get a photo. It was not close but at least he had a photo and could save what was left of his film for his real quest, the Yellow-headed Blackbirds.

It was a beautiful morning and the marsh was alive with many sounds. American Coots called and swam in and out of the reeds on either side of us. Periodically, we would hear the loud "cow-cow-cow" of the Pied-billed Grebe, the descending whinny of a Sora or the pumping "glunk-a-dunk" of the American Bittern. So many delightful sounds!

The full, bright yellow hood is the outstanding characteristic of a male Yellow-headed Blackbird.

Finally we heard a loud, raucous "squaaaaaaaaa" —a male Yellow-headed Blackbird! We had arrived at their location in the marsh and we could hear several around us and see them through the tall vegetation or flying in the distance, but nowhere did we find one posing for Jay to photograph. Jay now realized that he should have brought more film for he only had two or three shots left. We let the canoe come to rest and waited. With so many yellow-heads in the area, if we were just patient enough, surely one of them would come close enough for Jay to shoot his last few pictures. Some of them did come close but they seemed to be always behind the tall vegetation. We continued to wait. Sure enough, one male finally flew out of the marsh and landed in perfect view right in front of us.

Alas, Jay still was unable to get the photos he so desired. In many ways, this was the most cooperative Yellow-headed Blackbird that I have ever seen, for he was not just in front of us, he was sitting right on the front tip of the canoe just two feet from Jay and his camera! He stayed there almost a full minute: the perfect photo opportunity except for one small fact. You should recall that Jay had decided he only needed to bring his 400mm telephoto lens and none other. That lens had a minimum focal distance of over fifteen feet. Here was a bird at arms-length, sitting very patiently looking at us and Jay had no way to take a photo—the bird was simply much too close. Suddenly, the magic was broken and the bird flew back into the marsh. Jay raised his camera and snapped his last two shots as the bird disappeared. I suspect that they were only blurry images of a bird in hasty retreat.

Marsh Wrens nest in habitats favored by Yellow-headed Blackbirds.

I understood Jay's frustration, but for me it was a wonderful day capped by a very precious moment, that rare occasion when a wild bird allows so intimate an exchange. The photo opportunity may have been missed but the image of that one bird, sitting on the end of our canoe, will always be fresh in my mind.

24

Purple Martins

America's Largest Swallow

Purple Martins are one of the most beloved of all North American bird species. For centuries, Native Americans, especially in the East, attracted these birds to nest near them using hollow gourds, a practice that soon spread to European colonists. As early as 1712, there were published reports of Purple Martins coming to gourds in settlements in Florida and the Carolinas. People today still provide a wide variety of boxes and gourds to try to bring these attractive swallows to their yards. Such human fascination has been good for the martins but has, unfortunately, led to much misinformation and many misunderstandings about them.

Purple Martins are the largest species of swallows in North America and are the only swallows in which the male is entirely dark; the males of other swallow species have lighter underparts as do female Purple Martins. There are three races of this species in North America, and the largest populations are found in the race that occurs east of the Rockies. A second race is

Unlike most swallows, Purple Martins have distinctly different male and female plumages. The male, facing to the right, is all dark.

found in the Southwest down through Central Mexico, and the western race of Purple Martin is found in portions of the Pacific Northwest and in the western Rocky Mountains.

The information in most of the popular literature refers to the eastern race of Purple Martins, and in many ways, their behavior is different than that of martins in the West. It is often stated that these birds are colonial in their nesting habits, but it may simply be that many individual pairs tolerate nesting close to others rather than it being a true colony with its own colonial integrity and benefits.

It is instructive to compare Purple Martins to some other swallow species. Bank Swallows are colonial and the whole colony often comes into synchrony: egg-laying and hatching occur throughout the entire colony on nearly the same date. Rough-winged Swallows, on the other hand, nest in groups that may look like a colony but this is simply because they must use a very limited type of nesting habitat. Aside from nesting near one another, Rough-winged Swallow pairs act

Rough-winged Swallows nest close to each other, but no true colony is formed.

individually and pay little regard to neighboring pairs. In many respects, Purple Martins behave more like Rough-winged Swallows, but more detailed study is still needed to discern to a greater degree just how Purple Martins differ from other swallows.

In the West, Purple Martins are usually solitary, preferring old woodpecker holes. In fact, it is very difficult to get them to nest in artificial boxes or gourds and they are almost never found in suburban areas, as is the eastern race. After many failed attempts, some box designs were finally successful at attracting martins in some locations in Oregon many years ago. Most of these locations are near open water and not close to suburban or urban environments.

The eastern race of Purple Martins is entirely different in this respect. Not only do they come to yards and nest boxes, but there are currently no known eastern Purple Martins nesting in natural cavities; all known nesting pairs are in artificial structures. In fact, since 1900, there are only a small handful of records of eastern Purple Martins using natural cavities at all.

One of the reasons that Purple Martins may be so popular is their reputation for eating mosquitoes. However, this belief is incorrect, and numerous studies over the last 100 years

Photo by Barbara Gleason

A Purple Martin nest box from the East coast, this one in Virginia, shows the "apartment-style" houses used by martins in the East. This type of box is not taken by West coast martins, though. They prefer single smaller houses adjacent to one another.

have shown that Purple Martins very rarely eat mosquitoes. Claims of eating 2,000+ mosquitoes a day are not merely exaggeration but are complete falsehoods. The reason they don't eat mosquitoes is simple: mosquitoes generally fly near the ground, mostly at night, or at dawn and dusk. Purple Martins, on the other hand, feed high in the air and are most active during mid-day. Thus, a chance encounter between a martin and a mosquito is unlikely. Even when the two species may overlap, martins will usually ignore insects the size of mosquitoes, preferring larger insects such as damselflies, dragonflies and craneflies. Some beetles and other large flying insects are also eaten. So, Purple Martins are attractive and fun to have around but will do absolutely no mosquito control.

Instead, if you really want to help control mosquitoes, do what you can to encourage bats. Little brown bats are common and each individual will eat 400-600 or more mosquitoes every night.

To capture flying insects, Purple Martins fly at somewhat higher altitudes than do other swallows, typically from 100 to over 400 feet above the ground. Because they fly high and usually do not feed in groups, they are often overlooked by the casual observer. They seldom forage if temperatures are much below 50° or in heavy rain, as there are few flying insects available to feed upon in such conditions. Individual martins will sometimes go as long as 48 hours without food and pairs will usually huddle together in a cavity to help keep warm during such cool periods.

Purple Martins spend the winter in South America but arrive in the U.S. in spring to begin the breeding cycle. The first birds to arrive are the older, more experienced males, and they seek the best cavities and defend them from the later-arriving, younger males. When the females arrive over the next 5-10 days, the males begin calling to attract them. Females seem to be most interested in finding good nest cavities rather than being attracted to any particular male. Once a pair-bond is established between two birds, thc female begins nest construction in the cavity and the pair will raise their single brood.

As mentioned earlier, eastern Native Americans sometimes used hollowed-out gourds to attract martins. Even today, Purple Martins seem to prefer gourds (natural or artificial) over birdhouses. Gourds usually provide a deeper cavity which leads to greater nesting success and single gourds are often easier to defend than nests in apartment-style houses.

Most nest construction is done by the female although the male often starts the process by gathering a few small twigs. Both male and female continue to bring fresh green leaves to the nest, putting them in around the eggs. These leaves are removed once they have dried and become brittle. It is not known with any certainty why martins do this. Three to

A pole with a platform on top to attract nesting Ospreys, and boxes on the pole's side that have attracted Purple Martins. Western Martins prefer natural cavities for nesting.

six eggs is a typical clutch size for Purple Martins, although nests with single eggs, or as many as eight, have occasionally been reported. Both sexes of the pair will feed the young once the eggs have hatched. Soon after the young have fledged, the parents will take them away from the nest site. If they remain in the nest area, they are often harassed by other martins in the colony. This most often happens among the eastern populations that nest together in large numbers, and is less problematic for western birds that tend to be more solitary. Just 4-5 days after leaving the nest, the young begin to catch their own food and by 10 days after fledging, the parents often leave the young entirely on their own.

If you live where the eastern race of the Purple Martin flourishes, you may wish to put up a gourd or box to attract them. They can be entertaining to watch and their liquid, chattering songs and calls are a delight on a warm summer day. Attract them for your enjoyment and because you want to provide them with shelter and protection, but remember, that if you have a mosquito problem, attracting bats is a better solution than Purple Martins.

25

What's in a Name?

With binoculars at the ready, everyone in the group peers intently into the bush looking to find what is responsible for the movement of the leaves. "There he is," someone shouts and points. All eyes turn and are now focused on the Black-capped Chickadee, silently gleaning insects, spiders or their eggs from the surfaces of the leaves.

The scene above is a common one, and has within it, a common problem—a problem reflected throughout our language and our writing. A male Western Tanager is easy to distinguish from its mate, but unless the chickadee sings, we cannot tell by sight if it is male or female. Yet, we often say "he." This male bias universally common in our language and culture, and most of us repeat the bias without even thinking about it. It is not deliberate, and we all know that "he" is not necessarily male. But, I think we need to be aware of what we are doing and make a conscious effort to change if you or others are troubled or offended by it.

In the case of birds, this issue goes one step further for many species. The names we have given to many of our birds are, in reality, only appropriate for one sex. We have Black-throated Gray, Black-throated Blue and Black-throated Green Warblers, yet, in none of these species do females have black throats. The list goes on—Red-winged Blackbirds (only the male has red on the wings), Lazuli Bunting (only the male is azure-colored, from which the name, Lazuli is derived), Red Crossbill (females are brown streaked), Redhead (the female duck is brown), etc.

Florence Merriam Bailey

Aug 8, 1863—Sept. 22, 1948

Florence Merriam Bailey was an important American naturalist and one of the first women to write detailed studies of birds. Bailey wrote her first field guide on birds at the age of 26. She became the first woman associate member of American Ornithologists Union (AOU) in 1885 and was its first woman fellow in 1929. She also was the first woman to receive the AOU's Brewster Medal in 1931 for her book, Birds of New Mexico, *and was a founding member of the Audubon Society of DC and taught classes in ornithology.*

She married Vernon Bailey December 18, 1899, and together they studied and published information and life histories of birds and mammals of the West. A subspecies of the Mountain Chickadee, referred to as the California Mountain Chickadee, Pocile gambeli baileyae, was named in her honor in 1908.

This naming discrepancy is lamented by Florence Merriam (later Florence Merriam Bailey) in her book, *Birds Through An Opera-glass* (1889), as she talks about the Black-throated Blue Warbler:

Like other ladies, the little feathered brides have to bear their husbands' names, however inappropriate. What injustice! Here an innocent creature with an olive-green back and yellowish breast has to go about all her days known as the black-throated blue warbler, just because that happens to describe the dress of her spouse! The most she has in common with him is a white spot on her wings, and that does not come into the name at all. Talk about woman's wrongs! And the poor

Only the male Lazuli Bunting has the blue that gives the species its name.

> *little things cannot even apply to the legislature for a change of name!*
>
> *You do not blame them for nesting in the mountains and the seclusion of northern woods, to get away from the scientists who so ignore their individuality.*

There is a certain logic to this pattern of naming birds as males are often the most colorful and, since they sing and display, they are also the most easily seen. Historically, most birds were usually named by male observers, and I doubt they were deliberately trying to ignore females; it's merely a function of our cultural bias. In fact, some women were being honored. Lucy's Warbler was named to honor Lucy Hunter Baird (1848-1913), the daughter of Spencer Fullerton Baird, one-time secretary of the Smithsonian Institution. Virginia's Warbler was named by Baird to honor Mary Virginia Anderson, the wife of William Wallace Anderson, a U.S. Confederate Army surgeon who collected the first specimen of this species. Anna's Hummingbird was also named for a woman (Anna de Belle Massena, 1806-1896,

Duchess of Rivoli, France). Such names are few in the list of common names for North American birds; however, a few more examples may be found in birds' Latin names, *Athene cunicularia* (Burrowing Owl), for example, honors the Greek goddess Athene (or Athena).

In many cases, the names we use are not really appropriate for males throughout their entire year either. Look at the Scarlet Tanager, for example. The female is never scarlet in color, but for a majority of the year, neither is the male. From August through March, the male has a greenish-yellow body. He is scarlet only during the breeding season when he visits North America, away from his winter residence in Panama and South America.

Many names reflect this North American bias, as well as a sexual bias. We like to think of the many neo-tropical migrants as "our birds." Yet, many species spend more time away from North America than they do here. They might be better thought of as tropical species that simply come here to breed.

I don't have suggestions for how to change these long-standing names, nor do I advocate doing so. "Marsh Blackbird" would certainly apply equally to male and female Red-winged Blackbird but this would be a poor choice as it does nothing to distinguish it from Tricolored Blackbirds or Yellow-headed Blackbirds, both of which are also in the marsh. Instead of changing the names, I simply ask that we be aware of what we are doing and consider not using "he" until you are sure that what you are observing is really a male. Instead of saying "There he is," say "There it is," until it sings to prove it really is "he."

20

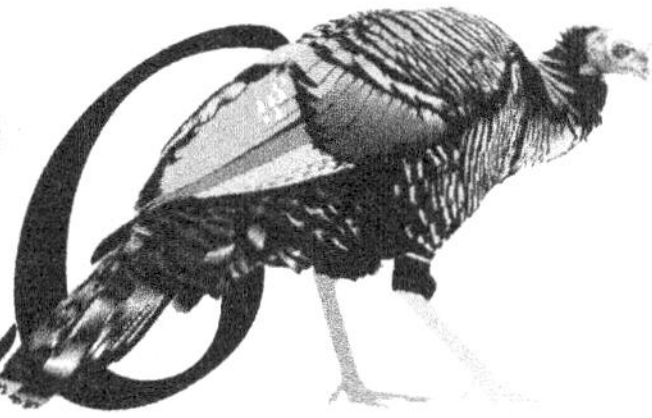

Gobbler in the Woods

Wild Turkeys are native North American birds, but they are not native to Oregon and sightings were once quite rare. Today this has changed and those of us who live in wooded or semi-wooded portions of the Willamette Valley have likely had some encounters with Wild Turkeys. If you live in the Pacific Northwest, perhaps you have seen a group of these birds wander through your neighborhood and wondered where they came from. And, if you are new to an area with turkeys, you might not know that the Wild Turkey is also a newcomer. How it got to Oregon, and particularly, to the Willamette Valley, is somewhat unclear. Certainly, there were deliberate introductions, but when and where is not well-documented for all populations.

Wild Turkeys are now found throughout most of the continental United States but their historic native range seems to have been along most of the east coast from New England in the north to Florida in the south. The western limits of its native range are not well-defined but appear to have been from Pennsylvania down through

Male Wild Turkey, showing his wattle, draped over his beak.

Missouri into southern Texas. There are some early records to suggest that turkeys may have extended their range as far north as southern Colorado and south into eastern Mexico.

When early Spanish explorers reached Mexico in the early 1500s, they found some of the people of eastern Mexico keeping turkeys as a domesticated animal. By the mid-1500s, these domesticated turkeys were brought to Europe and widely introduced there as the explorers and their crews returned. When the colonies were being established in eastern North America, early settlers brought turkeys with them. Wild Turkeys already were present, but the colonists released and maintained their domesticated stock as well. In the late 1600s there are accounts about how numerous turkeys were, but by the mid-1800s, the native birds were nearly extinct in eastern North America due to over-hunting. By the end of the nineteenth century, it is thought that only Florida, and possibly some remote areas in the Appalachians, had viable populations of Wild Turkeys in eastern North America.

Male Turkey in full display, trying to both attract females and intimidate males.

Reintroductions from wild populations were very successful during the twentieth century and turkeys are now flourishing and increasing in all of the lower 48 states.

Turkeys in Mexico may have ranged as far north as Colorado. There is little documented information but it is likely that these populations continued to flourish even as the eastern birds were disappearing, and there is some evidence to suggest that these birds continued to be hunted by the native people in the southwest and Mexico in much the same manner as had been done historically.

The Wild Turkey is a very large, and sometimes aggressive, bird. Where we live, in the south hills of Eugene, Oregon, turkeys frequently come through our yard. We have watched them chase squirrels, neighborhood cats and even deer.

Wild Turkeys were once nearly extirpated from their native Northeast habitats. Reintroductions allowed them to once again flourish in many parts of the country, including Oregon where they are not native.

A few years ago, a large buck with a full set of antlers was lying in our back yard resting when a group of turkeys came strutting through. The buck paid no attention to them, but one of the turkeys apparently wanted to feed where the buck was resting. We watched as the turkey pecked vigorously at the flank of the buck numerous times. Eventually, the buck got up and left the yard to avoid further harassment.

The plumage of Wild Turkeys is an iridescent dark black or brown that almost looks metallic in some light. The feathers are tipped with a rusty-brown color (especially in females). The race of turkeys that originated in Mexico and was brought to the U.S. via Europe, has white tips to their feathers, giving them a somewhat different appearance. It is from these white-tipped birds that domestic white turkeys are descended.

Wild Turkeys spend most of the day wandering through woodlands in search of food. A few insects or other

Wild Turkeys are able to fly into trees to roost at night when they are only 2–3 weeks old.

invertebrates may be eaten but the majority of the diet is plant material. Nuts and grains of all kinds are eaten along with a few grasses or other leafy vegetation in the summer. They also ingest large amounts of grit which is retained in the gizzard to help with grinding food. The gizzard is the rear chamber of a bird's stomach and is well-developed in seed-eating birds and especially so in turkeys. A turkey has no difficulty in pulverizing hard-shelled nuts such as acorns or, in the East, pecans and hickory nuts. Walnut shells are completely ground after about four hours in the gizzard. Some strange experiments were done on turkeys brought to Europe in the mid to late 1700s. One person reported that 12 steel needles were completely ground up by a turkey in 36 hours and another reported that steel tubing was flattened and partially rolled after 24 hours in a turkey gizzard.

The white muscle of the breast that so many people cherish for their Thanksgiving Day table, is an indication that this bird is not capable of long, sustained flights. Instead, it can make very powerful, quick bursts of flight. (Birds that are long distance migrants have red breast muscles, not white.) In turkeys, flight is most important as a means of escaping predators.

Rather than flying, turkeys usually move from place to place by using their powerful legs, the muscles of which make them strong, swift runners. They are even good swimmers, although swimming is most often observed in young, still flightless birds that are attempting to elude danger.

Turkeys nest on the ground but once the eggs have hatched, the parents do not directly feed the young, so chicks must follow the adults, learning what to eat by imitation and trial-and-error. At night, turkeys retreat into the forest where they fly up into the trees to roost on a branch. Even though several birds may be clustered together through the night, they are seldom observed by most of us.

In my neighborhood, Wild Turkeys were once unknown, but now, no one would be surprised by the presence of turkeys in their yard. They wander over a large area and flock sizes may vary in numbers each day. You may find one or two birds one day and then not see them again for several days. When they do return, you may see more or fewer than were present on their last visit. We often have a single bird wander through our yard and we have also had a flock with more than 30 individuals come to feed for an hour or so.

Turkeys are ground-foraging birds and feed upon whatever has been scattered on the ground including seed that may have fallen from overhanging bird feeders. We often have turkeys fly up onto our second-story deck and peck at the bird feeders that hang there. They can see the seeds through the transparent tube, but they peck at the tube, seemingly unaware that holes exist where seed can be obtained. Lately they have figured out that the suet feeders contain yummy goodies, too, so we try to discourage them from eating us out of house and home! They may also cause some damage to your garden; although they may not *eat* many of your plants, they can damage or dig them up by scratching and turning the ground in search of food. They can also pose an additional hazard as they have been known to chase pets and even people in some areas.

For you trivia-minded readers, you might like to know what to call a group of turkeys. You may have heard colorful terms for birds, such as "a flight of swallows," "a charm of finches" or "a wedge of swans," but the phrase for a gathering of turkeys is "a rafter of turkeys," ...unless it's November, in which case the phrase becomes "a death-row of turkeys."

27

Great Blue Heron

A large bird stealthily and methodically inches through shallow water along the river's edge on long, stilt-like legs. Its yellow eyes peer down along its dagger-like beak, staring intently into the water below. Suddenly, its outstretched neck thrusts its head downward into the water. A second later, the bird's head reemerges, a fish tightly held in its beak. A Great Blue Heron has taken the first of several fish it will capture today.

Great Blue Herons are the largest wading birds in North America, standing just over five feet tall and weighing up to 5.5 pounds. Great Blue Herons are found throughout most of the United States and southern Canada and along both coasts of Mexico as far south as Guatemala. They are equally at home searching for food in freshwater streams, lakes and marshes, or feeding along margins of saltwater bays and coastal streams. Sometimes, they are seen foraging in fields far from open water, where they hunt for mice and other mammals,

Great Blue Herons are familiar inhabitants of waterways in North America.

but people are probably most familiar with them along rivers, in ponds and wetlands.

The Great Blue Heron's favorite food is fish but they also eat amphibians, reptiles, aquatic invertebrates and even some small mammals and birds. The number of fish eaten depends on the sizes of the fish a bird catches, and sizes range from 2–12 inches long. Fish are not speared, as some might presume, given the shape of the beak; a heron captures fish between the upper and lower halves of the beak. Occasionally, a fish is speared by accident and this creates a problem for the heron. With a fish stuck on its beak, it cannot open its mouth to swallow it. It must first shake the fish off, then re-catch it between its mandibles. The heron manipulates the fish, sometimes with short tosses in the air, until the fish's head points toward the bird's throat. The fish is then swallowed whole and, as it slides down, the fins fold flat against the fish's body allowing it to be more easily swallowed. Swallowing a fish tail-first would cause the fins to expand and make swallowing very difficult. Fish-eating birds

Great Blue Herons sometime seek prey in deeper water.

know this trick and swallow their captured fish head first. (The Osprey is an exception, taking its fish to a perch and tearing it apart with its sharp beak, rather than swallowing it whole.) While fish are the main prey, some herons eat enough small mammals (mostly voles, but occasionally something larger) to account for 25-40% of their diet. Undigested parts, such as mammal fur, fish scales, exoskeletons of crayfish and other invertebrates, are coughed up as a pellet.

Herons hunt alone but are often seen near one or more other herons, and this is especially true during the non-breeding season. They wait patiently while looking into water, or they may slowly walk along, creating as little disturbance as possible so as not to scare away any fish.

The eyes of a Great Blue Heron are very well-adapted for night vision and herons may spend as much time foraging at night as they do during daylight. Many fish are less active at night, resting in shallows along the water's edge. The heron's night vision makes it easy to find them and the bird does not have to contend with bright daylight reflections on the water's surface.

All herons are well-adapted to looking down into water. If you get an opportunity to watch one closely, notice the position of its eyes: they are rotated slightly forward, to look beyond the beak as well as to the side, and slightly downward so that the head need not point down. This feature of herons is most exaggerated in the American Bittern, a heron restricted to marshes. In bitterns, the eyes are positioned looking downward, allowing the bird to focus on what is below while the beak points forward. In addition to seeking food, this serves another purpose: when the bird senses potential danger, it points its beak skyward. In so doing, its best field of view is now straight ahead, looking directly at the possible danger. The long slender neck is streaked and when its beak is pointed up, the bird blends in with the tall reeds and cattails of the marsh. Additionally, it sways back and forth like the nearby vegetation moving in the wind.

Herons are sometimes confused with cranes but the two kinds of birds are very different and not at all closely related. Cranes hunt in tall grass catching snakes, ground-dwelling birds and small mammals and when they fly, they hold their necks out straight in front of them. Herons in flight, however, tuck their head back in toward the body, the neck in a tight S-configuration. Other behavioral differences can be observed during the breeding season: cranes nest on the ground and herons make large stick nests in tall trees. Being ground-nesters, cranes have need of well-developed legs upon hatching. In herons, the legs are not well-developed until close to fledging time, which helps to

The Great Blue Heron is North America's largest heron.

prevent active youngsters from wandering and falling out of nests high in trees. Additionally, cranes usually mate for life, whereas herons must find a new mate each year.

Watching a large, long-legged bird land in a tree may look awkward to us, but herons do so, both to rest and to build nests in which to raise their young. Landing in a tree is something that you will never see a crane do.

Occasionally, herons nest in single pairs but most often, Great Blue Herons nest in loose colonies of 3-40 nests, and some nest colonies have been known to contain over 700 nests. Nests may be used by different pairs each year but the colony site is often used for many years, with some colonies having had active nests for 50–70 years. Colonies are abandoned when the surrounding habitat is destroyed or if heavy predation occurs. In fact, in some places, the return of breeding Bald Eagles has been detrimental to Great Blue Heron breeding colonies. Eagles will sometimes steal sticks from heron nests before the herons have begun to breed, and once breeding is underway, the eagles may capture young herons in the nest or even kill an adult. Such harassment usually causes the herons to move elsewhere and establish a new nest colony.

One Great Blue Heron nest with two young birds in it, some 50-70 feet up high in a Cottonwood tree.

Sometime they move to a new area altogether and sometimes they simply move to another grove of trees even less than a mile away. Either way, it can disrupt that year's breeding cycle. In addition to eagles, Great Horned Owls are also a major predator at heron nests.

Eggs are usually laid by late February or early March in western Oregon, and most young herons in such nests are ready to leave the nests by the end of May or early June. Once the fledglings can fly, they take short flights but return to the nest to be fed for a few more weeks, as the parents will not feed them anywhere except at the nest.

The shape of a heron's neck is a gentle S-shape when the bird is standing relaxed or when it is ready to strike, although it may stretch out its neck when it focuses on potential prey in the water below. While keeping its head still and focused on its prey, the bird very slowly advances forward so that the body moves closer to the head, reforming the S-configuration of the neck. Now the bird is ready to pull the head back and swiftly stab it forward, hoping to snatch its prey in the water. By forming the S in its

A drawing showing how an atlatl assists in striking one's prey.

neck, the bird has greater striking distance and power.

Additionally, you will notice that the neck looks like it has a kink in it. If you could look at the skeleton, you would see that the neck vertebrae fit together at that point at an angle, as if the neck had been broken. In fact, I have seen skeletons in museums where this assumed "defect" was "corrected" and made to look smooth. But, this neck shape is not a defect and, combined with the associated muscles, it actually gives the bird just a bit more power when throwing its head forward to seize prey. It is similar to the advantage given someone throwing a spear if they use an atlatl (spear-thrower).

There is another behavior that Great Blue Herons occasionally do, especially on warm sunny days and if you watch them often enough, you may finally be lucky enough to observe it yourself. They will sometimes do what is called a "spread-wing posture" where the wings are slightly drooped and tipped back, exposing the underside of the wings to the sun. This is a form of sun-bathing as they open their wings to gather the heat of the sun. It is possible that the heat helps the birds to rid themselves of parasites. While this has not been verified in herons, there are other kinds of birds who use the heat of the sun to cause enough discomfort to parasites that they will dislodge from the feathers or skin and either drop off of the bird or at least become easier for the bird to remove. It would be interesting to know if that is what the heron is also attempting to do by this behavior.

Even people who are not bird-enthusiasts are excited the first time they see such a large bird nearby. Great Blue Herons' size and gracefulness are impressive and they serve as a symbol of clean, free-flowing waters, even within the confines of urban environments. They are magnificent birds with which to share our world.

28 A Green Heron Goes Fishing

The Bolsa Chica Ecological Reserve, along southern California's coast, is a wonderful place to visit, especially for those interested in birds. It is not a "wilderness" experience, since Highway 101 runs along the western edge of the reserve and oil derricks dot much of its southeastern landscape, but this salt marsh offers safe haven to many bird species and the boardwalks and trails through and around the perimeter offer birders and photographers a great opportunity to see many of these birds, often at close range. Terns, skimmers, pelicans, many kinds of shorebirds, herons and egrets—all are present and easily seen. I have had many wonderful experiences here but this story concerns a Green Heron.

A small member of the heron family, the Green Heron is about the size of a crow, and it often tends to be somewhat shy and secretive. Occasionally, though, it stands in the open at close range, seemingly unconcerned about people passing within a few feet of where it is hunting. I have seen them close on several occasions but one time was especially rewarding.

This Green Heron has just caught lunch!

While birding at Bolsa Chica Ecological Reserve, we came around a bend in the trail along the edge of open water and before us at the foot of a small bank, was an immature Green Heron. It paid us no attention: it was focused on looking for small fish. It speared the water with its dagger-like beak, came up with a fish, which it promptly swallowed...then intently stared at the water once again. It was completely unconcerned that we were so close—the perfect photo opportunity. I steadied the camera with its 400mm lens and carefully focused on the heron. Sure enough, a fish soon caught its attention. It very slowly and deliberately took a couple of steps along the shore, moving closer to the fish. Ever so slowly it moved until the tip of its beak was just an inch above the water's surface. Suddenly, out and down darted the bird's head and neck again, in a movement so fast that my camera recorded just a blur. It again had a small fish in its beak. Then, in one quick motion, the heron flipped the fish around and re-caught it so that the fish's head was pointing inward. The heron's head tipped up, and down the fish went, swallowed whole, head-first. (See Chapter 27, for

An adult Green Heron patiently stands in the dry grass along the side of a stream.

details about how herons capture and swallow fish.) All of this was a real treat to watch and better yet, I had it all on film. But this heron had one more treat for us that day.

Not everyone who visits Bolsa Chica and other sites in our natural world has full respect for the landscape and its residents, as is evidenced by some of the litter along the water's edge. It is not excessive but any amount can be an intrusion. On this day, there was a yellow cracker lying among the small rocks at the edge of the water. This attracted the attention of the Green Heron. It picked up the cracker and very delicately laid it on the surface of the water and then stood very still and waited. The color of the cracker acted as a lure to the fish, and they approach to examine it as a possible food source. As fish come to investigate, they may become dinner to a patiently waiting heron and we were fortunate enough to watch this marvelous behavior.

This Green Heron has set a cracker onto the water's surface, and now waits for an unsuspecting fish to come investigate, called "baiting."

This "baiting" behavior has been well described but this was the first time that I had seen it first hand. What a rare treat from a very cooperative bird!

On another occasion, I watched a Green Heron fishing as I visited a local Eugene pond where Green Herons are frequently found. This individual was an adult standing on a log in the pond, perhaps only 30 feet from me. It was intently focused on something in the water. It slowly crept forward with the tip of its beak near the water's surface. Suddenly, the head and neck darted forward into the water and in an instant, the heron plunged into the water—completely immersing itself. It looked clumsy, as if the heron had slipped off of the log inadvertently. In reality, the fish it sought was likely deeper than just below the surface. It's not typical of most herons to plunge their whole body into the water, but this technique is sometimes used by Green Herons. Had this action not been deliberate, the heron might well have missed his prey, but he climbed back onto the log and had a small fish neatly caught between his mandibles. To my delight, this same hunting strategy was used again a few moments later, and again successfully.

29

Barn Owl: Evening Ghost

As its common name suggests, Barn Owls may be found associated with barns and other structures in agricultural regions and they are often highly valued by orchardists and farmers world-wide. These birds are also sometimes found in open urban expanses, such as large parks.

There are few species of land birds as widely distributed as Barn Owls, which are found on all continents except Antarctica. Over 28 different races are known, and although they are members of the same species, these different races show considerable variation in size and color. Barn Owls in the Galapagos Islands, for example, are only about half the size of North American Barn Owls, and our owls tend to be much whiter than those in Europe.

Many readers may never have seen a Barn Owl, even though they are relatively common. Very nocturnal, they often do not become active until well after sunset. They often roost and nest in human structures such as barns or silos, hence their common name, but in Europe, some use steeples of old churches and are sometimes referred to as "Church Owls." Because of their heart-shaped faces, they have sometimes been called "Monkey-faced Owls," and their plumage, white on the underside, earns them the nickname of "Ghost Owl" or "White Owl." The white has also led some people to mistake them for Snowy Owls.

Most data indicates that Barn Owls are non-migratory, remaining in a given area year-round, even in their most northern habitats. There are scattered reports in the northern United States, of movements of Barn Owls, but it is unclear if these reports represent true migration. They feed mostly on small rodents, primarily voles, which is why they are well-liked by many farmers and orchardists. A large population of voles can devastate a grain crop, so attracting Barn Owls is one of the best ways to control a rodent population. Old wooden barns are generally more conducive to the needs of nesting Barn Owls, as they offer more rafters and lofts which are often lacking in newer, metal constructions. This lack of nesting sites, along with habitat destruction and some agricultural practice changes, have caused a decline in most populations of Barn Owls. To help offset this decline, farmers can build large nest boxes to attract these beneficial owls. Farmers and vineyards in Oregon and Central California and rice growers in Malaysia have all had success in rodent control by putting up nest boxes for Barn Owls. However, in areas of monoculture and lack of hedgerows or fallow fields, the owls probably won't do well. You can obtain plans for building such boxes from Cascades Raptor Center in Eugene, Oregon at: http://www.eraptors.org/nestbox-barnowl.htm

Shrews, bats and some small birds are occasionally eaten by Barn Owls but they rarely eat reptiles or amphibians. They have sometimes been called "Gopher Owls," but adult gophers are too large for Barn Owls to prey on, although baby gophers might be taken.

Barn Owls are easily recognized by their distinctive, heart-shaped faces, and relatively small, all-dark eyes. Male and female look alike in most owls but male Barn Owls have much whiter bellies and underparts than do females, which are tawny-colored with many small, dark spots. The more spots she has, the more resistant the female seems to be to lice, parasitic flies and feather parasites. Such densely-spotted females are also much more attractive to males. Mated pairs often preen each other and the more spots on the underside of the female, the more frequently her mate will preen her, thus paying her more attention and creating a stronger pair-bond. He will also give the young more attention once her eggs hatch.

The long legs are a notable characteristic of Barn Owls as these enable them to more easily capture prey in tall grass or weedy fields. Some people are surprised to learn that it is not necessary for Barn Owls to see their prey for successful capture. Experiments have shown that Barn Owls can accurately locate and capture live prey in total darkness. Their ability to hear is so keen and well-focused that the owl can easily hear the movements of prey in the grass or other ground cover.

Silent flight is a characteristic of most owls, not just the Barn Owl, and extremely soft feathers over the wing help an owl fly silently. It is often assumed that silent flight enables an owl to sneak up on its prey without the prey hearing them coming. However, day-hunting raptors, including day-hunting species of owls, do not fly silently and are quite capable of capturing prey efficiently. Thus, a more likely reason for silent flight is that it enables the owl to listen for sounds made by possible prey without the competing sounds of its *own* wingbeats.

Barn Owls are short-lived, and up to 75% may die before

(Photo used with permission of Cascades Raptor Center.)

The Barn Owl has sometimes been called "Ghost Bird."

reaching one year of age; most don't live longer than two years in the wild, but a few old birds are known. One study, conducted over a 20 year period in Utah, found an owl that lived to be 8 years old, and the oldest known Barn Owl lived 34 years. Even though they seldom live more than two years, Barn Owls can be quite prolific, especially in years when the prey-base is high. Each female lays 4-9 eggs per brood and 2 broods are not uncommon in southern portions of their range or in good rodent years.

Raising a brood of Barn Owls can be quite a challenge for the parent birds. Eggs are laid at 2–3 day intervals, which can mean that the first egg and the last egg can be as much as 20 or more days apart. The female Barn Owl begins incubating when the first egg is laid, thus hatching is also spread over a 20+ day period. Imagine tying to feed a brood where the youngest has just hatched and the oldest is nearly ready to leave the nest, though!

The white underside of a Barn Owl may be easily seen in moonlight if one happens to fly in front of you. This flash of white can be an unexpected surprise, but it is likely that a more eerie encounter would be to only hear a Barn Owl's voice. They do not have a deep hoot like Great Horned Owls do, and they don't softly whistle or trill like Screech-Owls. Instead, they have a prolonged hissing-scream. Should you hear one over your shoulder at night, it is a sound that will certainly get your attention, and may send chills up your spine.

If you live in a suburban neighborhood, you will likely need to travel to a rural, open area to find Barn Owls, but you usually don't need to go far. Those who live in open, non-wooded regions might want to consider finding an appropriate location for putting up a nest box. You may be well-rewarded by having the opportunity to observe these beautiful owls, and know that any population of small rodents will be kept in check by an excellent natural control.

30

A Noisy, But Lovable Jay

If you live in the Willamette Valley of Oregon, you have likely watched them streak through your neighborhood: medium-sized, blue-colored birds making loud, raucous calls as they fly. You probably know that they are jays, and while many people call them "Blue Jays," they are correctly called Western Scrub-Jays. The real Blue Jay lives in the eastern United States, and only occasionally comes to Oregon in the winter.

Two kinds of jays are common to Oregon's Willamette Valley: the jay with a crest on a dark blue head is the Steller's Jay, common in coniferous forests and woodlands. The subject of this chapter is the other common jay, the Western Scrub-Jay, a non-crested member of the family found in oak woodlands, open countryside and suburban neighborhoods.

The Western Scrub-Jay is non-migratory and lives year-round in the western United States and Mexico. Young birds may wander but generally a pair claims a territory and defends

Western Scrub-Jays often show little fear of humans, especially in places where food is offered.

it throughout the year. Once mated, male and female typically form a life-long bond. As breeding season nears, the male usually remains near the female, and although a bird may feed its mate any time of year, mate-feeding is most common during the breeding season. Either sex may beg for food by making soft begging calls and quivering its wings, similar to the action of baby birds begging food. In adult birds, begging helps maintain and strengthen the pair-bond.

New pairs most often form only when one member of an existing pair dies. Unmated birds wandering in an area may then find an opportunity to gain a mate. Sometimes young birds, hatched in the spring, form a pair relationship that lasts through the winter, but many such pairings break down as breeding season approaches.

By mid-March many Western Scrub-Jays begin breeding and either sex may select the nest site. If the other bird accepts this site, nest construction begins and construction is shared equally. You may see jays go swooping by carrying small twigs. Such twigs are woven together forming a cup, which will then be lined with soft plant fibers and, often, mammal hair. Working together, the nest takes 1 to 2 weeks to complete. The female then lays her clutch of 4-6 eggs, one each day until the clutch is complete. She does all of the incubation and the male provides her with most of her food.

Western Scrub-Jays have one brood per year but may lay a second clutch if either the first nest, first clutch of eggs, or first nestlings are destroyed. The eggs hatch after 17-18 days of incubation and the young fledge when 20 days old. Young birds remain with the parents for at least two more months, well after gaining a measure of independence, but parents sometimes feed fledglings as old as 3 months of age.

Fascinating behaviors can be observed if you take time to watch these birds. Western Scrub-Jays can live up to 15 years and along with other jays and crows, are among the most intelligent of birds, so you may observe behaviors not found in most other birds.

Many birds store food in small caches (or storage locations), but jays and crows are especially adept at this. Scrub-Jays frequently bury food items, such as nuts and seeds, and sometimes store items under stones. Sometimes they will bury insects or scraps of human food they find. Perishable items are usually recovered quickly but nuts and seeds may be stored for months. Jays are known to have recovered nuts they buried more than 250 days earlier. Jays seem to know what they've cached, as well as where, and choose preferred foods first. Most items are recovered, but the habit of burying food may play an important role in seed dispersal of some oaks, and in the Southwest, of pinyon pines.

More remarkable are the behaviors that have been observed when Western Scrub-Jays are in the process of burying food. They may watch as other jays, crows or squirrels bury food and later go back to steal those items. If they discover that they are being watched when caching food, they may stop and simply wait until the intruder leaves. If the intruder does

The Western Scrub-Jay is a common backyard visitor in western Oregon.

not leave quickly, the jay may bury its food, but as soon as the intruder leaves, the jay returns and retrieves the buried item, moving it elsewhere so as not to reveal its location to potential competitors.

Jays are also excellent observers. If one jay eats something bad tasting and spits it out, other jays observing this behavior will avoid similar food without trying it.

Western Scrub-Jays are opportunists, eating a wide range of foods. Acorns and other large seeds are consumed in large quantity but insects and insect larvae are also eaten, and jays may capture small snakes or take eggs or small nestlings from other birds' nests. Jays usually smack snakes and caterpillars against a branch to kill them and caterpillars with long body hairs are also knocked on branches to remove the hairs. Scrub-Jays have been seen to land on the backs of deer where they remove parasites from the deer's skin. The deer usually make no attempt to discourage or remove the jays.

Western Scrub-Jays easily coexist in human environments. Although noisy at times, these intelligent birds can become welcome neighbors. Both Western Scrub-Jays and Steller's Jays can be attracted to suet or large seeds, and Scrub-Jays seem to be especially fond of peanuts (whole, unsalted, please). Feeding peanuts can become expensive, though, and you and your neighbors will find them buried in various locations around the neighborhood. Some people don't like jays, and if you are one of them, I would suggest giving them a second chance, as they are beautiful, smart and fascinating.

3

Both Hated... and Beloved

The European Starling is a beloved bird throughout much of Europe. In parts of Britain, its return each Spring is hailed as one of nature's wonders and current population declines in Europe are lamented. Yet, European Starlings may be one of the most disliked bird species in North America.

Introduced from Europe, the European Starling has probably played a significant role in the decline of some of our native birds, but we often ignore and misunderstand them. Even their name is often misunderstood. They are known to science by the Latin *Sturnus vulgaris*. For many of us, "vulgar" has a negative meaning, but in fact, it simply means common, and this is the common starling of Europe. *Sturnus* and starling mean the same thing: little star. It is suggested that the reference to star most likely comes from their silhouette in flight, as their long pointed beak, outstretched, triangle-shaped wings, and short tail somewhat resembles a star shape.

Starlings belong to a large, Old World family that also includes mynas, and, unlike the starling we are most familiar with, some starling species have spectacular plumage. In Europe, Africa and southeast Asia, 114 species of starlings exist but the family name, starling, comes from the plain-colored European Starling, the same species that was introduced into North America.

Hotspur:

He said he would not ransom Mortimer;
Forbad my tongue to speak of Mortimer;
But I will find him when he lies asleep,
And in his ear I'll holla 'Mortimer!'
Nay, I'll have a starling shall be
taught to speak
Nothing but 'Mortimer,' and give it him
To keep his anger still in motion.

—William Shakespeare:
Henry IV (part I), Act I Scene III

European Starlings are well known for their vocal ability, mimicking calls and songs of many other birds, often quite accurately. Mozart had a pet starling that he was very fond of, which amused him by being able to sing a short selection of his music. Mimicry of human speech by starlings was reported in the works of Aristotle, Pliny and the ancient Romans. Shakespeare also knew of this ability, for in Henry IV (Part I), the character Hotspur threatens to train a starling to say the forbidden name "Mortimer" just to annoy the king. In fact, it is this single Shakespearian reference that is indirectly responsible for the successful introduction of starlings into North America.

In the late 1800s, many species of European birds, including European Starlings, were introduced in the United States and Canada, but most of these introductions failed. Among the places European Starlings were released was Portland, Oregon, in 1889 and 1892, by the Portland Song Bird Club. Similar groups around the country also brought birds from Europe. The reason for the introductions were varied, but often it was immigrant groups that wanted to have some of the birds that they were familiar with in their homeland around them. By 1902, all of the starlings released in Portland had disappeared, as did most such birds released around the country. But, in 1890, Eugene Scheiffelin, of the American Acclimatization Society, released 60 starlings in New York City's Central Park, followed by 40 more the next year. His stated purpose was to introduce into the United States every bird mentioned in Shakespeare, and that single reference in Henry IV was the reason he imported and released the starlings. The birds from those two introductions succeeded where no others had before, and in less than a century, European Starlings had spread across the continent. They reached the far west, becoming common in western Oregon by the early 1960s. The North American population is now estimated to be over 200 million, approximately one third of the world's population.

Starlings are very aggressive birds when it comes to nest competition. They nest in cavities and will aggressively drive out other native cavity-nesters including birds larger than themselves. They compete with bluebirds, swallows, woodpeckers and will even drive Wood Ducks out of boxes or cavities. It is suggested that both the Eastern and Western Bluebird do not fare well in competition with starlings. This aggression toward native birds is probably what has been the cause of so much loathing on the part of birders, and some would like to see starlings destroyed. But, for all of the suggestion of harm to native birds, there is little hard scientific evidence to support this idea, except for a few species, such as sapsuckers and some bluebird populations. It is possible that the harm to many species is isolated and not widespread and it is also possible that adequate studies have simply not been done and the harm may be as significant as anecdotal evidence suggests. At this point, we really don't know where the truth lies.

We do know that human activity plays a large role in the success of the starling in North America. Forests and wild natural habitats are not conducive to starling success. This

European Starlings are adored in most of their native European habitats, but much disliked as an invasive species in North America.

is probably why most early introductions failed: there were simply no suitable places for starlings to live. By the time the last 100 birds were released, the region was much more urbanized and starlings found conditions under which they would thrive. Starlings are more tolerant of human activity than many native cavity-nesting species, so they adapted to nesting in cavities provided by buildings. As the human population of North America increased and European influence spread across the country, landscapes were altered, favoring starlings, often at the expense of native species.

Despite their bad reputation, starlings may actually be beneficial in some circumstances. They can aid some agricultural operations because they devour large quantities of snails, cutworms, beetles and other invertebrates that destroy domestic crops. They can also assist in our desire to have nicely manicured lawns that are sought after by so many homeowners. One of the pests that can be detrimental to lawns is the larvae of craneflies, specifically, the European cranefly, *Tipula paludosa*. This invader from Europe arrived in the Pacific Northwest sometime in the mid-1960s and has become a pest on lawns and turf grasses. The adult is often mistaken for a giant mosquito and is sometimes called a "mosquito-hawk," in the mistaken belief that it eats mosquitoes. But it is completely harmless and does not bite, sting or even eat as an adult. The larvae are found in lawns where they feed on the roots of the grasses. (Native craneflies in the Pacific Northwest are not found on lawns and are not considered to be pests.) The larvae of the European cranefly is eagerly sought after by Starlings who are adept at obtaining them from our lawns.

The cranefly larvae are usually just out of reach of most of our native birds. But, the starling has an advantage that helps it obtain the cranefly larvae. In most birds, the muscles that close the beak are stronger than the muscles that open the beak. In the European Starling, just the opposite is true. The starling pokes the tip of its beak into the hole of a cranefly larvae and forcibly opens its beak, expanding the top of the hole and allowing the bird to reach a little bit deeper. If necessary, it will repeat this process until it can reach the larvae. Meadowlarks use a similar behavior to reach invertebrates in soft soil and orioles probe into fruit and open the beak to expand a hole and gain access to more of the fruit.

However, in large numbers, starlings can be problematic, and in the fall and winter, they compete with native birds for wild fruits and grains. If certain foods become limited in their chosen habitat, they easily switch to another available food source, and some native species are less adaptable.

While I do all I can to personally discourage starlings from feeding at my bird feeders, from a biological perspective, they are very interesting birds. During the breeding season, males bring plants such as yarrow to the nest cavity and

Yarrow is often collected and woven into the nests of European Starlings.

weave them into the nest. Chemicals in these plants stimulate the immune systems of the hatchlings, increasing their likelihood of surviving.

As fall approaches, starlings molt and their new buffy-edged feathers give them a speckled appearance. As the seasons progress, the tips of these body feathers wear away until by spring, they look shiny and iridescent. Thus, the shiny breeding plumage is the result of old, worn feathers. Males and females look nearly identical but there are subtle differences. The base of the male's bill is pale blue and the base of the female's bill is pale pink, although this can only be seen at close range and even then, rarely. Recent research shows that the iridescent throat of the male looks different than that of the female when seen in ultraviolet light, which is easily seen by birds but not visible to humans.

I'm afraid that the European Starling is here to stay, but I still advocate doing all that you can to discourage them from being attracted to your yard. Avoid putting up nest boxes that will attract starlings. They cannot enter a hole smaller than 1.5 inches but this will also keep out some native cavity-nesters. Starlings do not eat seeds, so they are not attracted to your feeder (unless the seed becomes infested with insects) but they will come to suet feeders. A dome over a suet feeder may keep them out or you can use the type of suet feeder that is enclosed except for an open mesh on the bottom. Some such feeders have a roof-like cover, which further discourages starlings. Starlings don't like to feed hanging from below such a feeder but chickadees, woodpeckers, bushtits and other birds will easily do so.

European Starlings may not be your favorite birds, but like all else in nature, they are not really "evil" and do have some interesting qualities. They are in the United States because humans brought them to this country and it is we who have altered the landscape to favor their proliferation. They will undoubtedly be with us for a very long time so let's do what we can to minimize their impact on native species and appreciate the many interesting qualities from which we can learn.

Black-capped Chickadee, shown feeding at a suet feeder that requires the bird to cling to the feeder up-side-down.

32

"Tiger" of the Sky

The Great Horned Owl is a large owl, about 22 inches from head to tail, and ranges from the Arctic to northern Central America, and in a few high elevation portions of South America. These birds inhabit dense forests to grasslands and deserts. They are found throughout North America except at very high elevation or in urban areas, but are often common in woods in, or surrounding, suburban areas.

A non-migratory species, the Great Horned Owl is present and common year-round. Our heaviest resident owl (females average about 3 ½ lbs.), they are sometimes referred to as "hoot owls" because of the deep hooting call. The male's voice is the deeper voice, even though the female is always larger than the male, as is true for most raptors.

Great Horned Owls are formidable predators and will eat almost anything. Mammals are the primary prey, and can range in size from

(Photo used with permission of Cascades Raptor Center.)

Great Horned Owls frequently prey on skunks.

mice and squirrels to animals as heavy as 20 lbs. Rabbits, raccoons, skunks, an occasional house cat, and even porcupines are considered prey. Skunks are frequently eaten, and because (like most birds) owls have no sense of smell, they are not bothered by the odor. Great Horned Owls have large, very powerful feet, and while capable of catching and killing a skunk, or similar-sized mammal, they cannot carry something that heavy. Prey heavier than 4 to 4 ½ lbs. must be eaten or torn apart where it is caught because an owl cannot easily carry anything much heavier, just under twice the bird's weight. Birds are also taken by Great Horned Owls, from small songbirds and small owls (like Western Screech-Owl), to birds as large as herons and large geese. They sometimes take coots, ducks or other waterfowl sleeping on open water during the night. Interestingly, domestic chickens and ducks are almost never taken. When this does occur, it is usually a hungry juvenile owl that has not been successful at capturing wild prey. Most Great Horned Owls prey on nearly anything that has the potential to be food, and they also will eat earthworms, grasshoppers, crayfish, amphibians, reptiles and even scorpions, giving them the most diverse diet of any North American raptor.

Great Horned Owls range across North, Central and South America.

Large prey like adult skunks are often cached, or stored on the ground. The owl will return to this cache and continue to feed upon it. During cold winters nights, the carcass may freeze and make it nearly impossible for the owl to tear off more flesh. In such cases, the owl may sit on it "incubating," or warming it, until it has thawed sufficiently to be eaten.

Hunting activity begins at twilight and continues to a lesser degree through the night. Sometimes, Great Horned Owls need to hunt during the day, particularly when there are young to feed.

"Great Horned," refers to the long feather tufts atop the head and face, which are sometimes referred to as "ear tufts," a misleading term because they have nothing to do with ears. These feather tufts, along with the facial coloration, can help to create a cryptic pattern that can conceal an owl roosting against the trunk of a tree. The tufts are also visible at night when the owl is silhouetted and the birds can raise and lower them in silent communication with a mate. The real ears of an owl are on each side of its head and, as with all other birds, the ears have no external structure.

People are often surprised to learn that Great Horned Owls in North America begin breeding in mid-winter, from mid-December to early February. Like other owls, they build no nest, but use an existing structure, like a cliff ledge or an old hawk or crow nest. The female lays 1–4 eggs, although typically only two young are raised in a season. She incubates and tends the young and the male brings most of the food, including what the female needs, until the young are close to fledging, unless what the male brings is insufficient. Great Horned Owls are among the most long-lived of owls, up to 28 or more years, so each pair can raise many young in their lifetime. Male–female pair-bonds are known to last at least 5 years and probably are life bonds in most instances. Both birds of the pair stay in the same area throughout the year, however it is not known how much interaction occurs between them during the non-breeding season. A pair of owls defends their territory vigorously from other owls, but hawks and eagles are completely ignored, even if nesting just a few feet away.

The Great Horned Owl's nighttime habits, beginning in the twilight hours, along with their silent flight, allow these birds to go mostly undetected by casual observers, despite it being one of our most common raptors. Any wooded area is a likely place for Great Horned Owls, where they may sit silently each night, watching for hunting opportunities. You may hear owls hoot or be lucky enough to see one fly past, or you might find one by noticing where robins or crows are gathered, sounding off loudly during the day. Their commotion and din is a way to try to drive off any owl (or hawk) that they have discovered during the day.

This is a formidable and impressive bird. If you ever have the opportunity to see one up close at a wildlife center, zoo, or elsewhere, look at its large feet and imagine the strength and power of which this bird is capable. You will leave much impressed and with great respect for this "Tiger of the Sky."

33

Who's At My Feeder?

When one acquires a reputation as a local bird expert, it seems that everyone has a question or story to pass along, sometimes delaying me even when grocery shopping. But I don't really mind...

I come into contact with many different people and many of them have very interesting things to say about their observations, or questions that may never have occurred to me. Many times I have had to say, "That's a good question, but I don't know the answer. Here is my best guess..." Of course, my curiosity is then aroused and I have to do some searching in the ornithological literature to try to find how accurate my guess really was. Often, a question seems, to me, self-explanatory, but the person asking it may know little about birds, and is genuinely curious about what she or he has seen. As time goes on, I have found myself answering the same questions over and over, and I need to remind myself that what may be old to me may be quite fresh and exciting to the person who called or stopped me on the street.

Some questions, or variations thereof, are repeated often. Each spring, someone calls concerning a bird in their yard that they have never seen before. Often, it is a fairly common bird, but may not be a species that the caller has ever seen or heard about. The Western Tanager is a good example. As they arrive each spring, a few of these colorful birds turn up in the yards of people who do not know birds and have never seen a tanager before. The male Western Tanager is certainly a glorious and colorful bird, but maybe especially so when seen by the non-birder who didn't know such a beautiful bird even existed locally. Occasionally, tanagers arrive in large numbers and are seemingly everywhere, foraging in many home yards before moving on to their breeding habitat. Once, a woman called me and was very excited. She said, "There are at least ten parrots in my yard! Do you think they could have escaped from someplace?" I questioned her about the birds and had her give me some details. She admitted that she really didn't know anything about birds, she just assumed that

The bright colors of the male Western Tanager caused some non-birders to think that they were parrots.

these were parrots because they we so brightly colored. Eventually, she gave me a reasonably good description of a male Western Tanager, and indeed, that year there had been many reports of tanagers from all around town. She called back a couple of days later to tell me that she had gone to the bookstore to purchase one of the field guides that I had recommended. She said that she had no idea that there were so many different kinds of birds and many of them listed as common right here in our own area. I told her about the field trips offered by the local Audubon Society and suggested that joining one of these trips would be a good way to begin learning about what was here. She said that she would try that, because if there other birds around just as colorful as these tanagers, then it might be fun to learn about them. That is certainly a very satisfying kind of phone call to have been able to respond to, and it is a good feeling to know that one's efforts can help another better appreciate the natural world. I always hope that my answers to questions might also inspire the caller to want to enjoy the natural world more and to take efforts to help protect birds and their habitats.

There is, however, that occasional caller that one really wishes would find another "expert" to address their question. It's hard to deal with the person who already has their mind made up and really just wants you to confirm that they are correct...especially when they aren't! Such an individual generally does not want to hear the truth if it contradicts their own preconceived answer. I remember the woman who called to ask what kind of bird was in her yard. "It looks exactly like a robin but it's twice as big." When I explained that size may sometimes be difficult to assess and if it otherwise looks and acts like a robin, then it probably is a robin. She quickly assured me that she "...knew what robins were and that this bird is not a robin and was definitely at least twice the size of a robin. So what is it?" I tried to think of what else she could have seen. I suggested Varied Thrush and even Northern Flicker and cautiously probed for other details. I was informed that it was not any of these birds and must be a new kind of robin. No other answer was acceptable and she was clearly not happy that I didn't tell her she had discovered a new bird. "Why can't anyone make a book that shows what's really here?" I try to be as diplomatic as I can but I can't always satisfy everyone.

Some callers provide me many good clues so that I can provide an answer, but others just demand an answer without giving me much information to work with. I once had a call from a woman who wanted to know, "What kind of bird did I have in my yard? It's not in any of the books." I asked what it looked like and, rather indignantly, she responded, "How should I know. That was ten years ago and I don't remember what it looked like but it isn't in any of the books, so what could it be?" I explained that I couldn't give her an answer if she couldn't provide me with any details. "I thought that you were supposed to be a bird expert!" she responded and then hung up.

One spring, another gentleman called and asked, "Where are all of the western cardinals this year?" Red is the first thing that comes to mind when I think of cardinals, so I tried to think of what red birds he could have been thinking of. It certainly wasn't a Northern Cardinal as that most certainly is not an Oregon species. Perhaps House Finch or Purple Finch? Maybe it was the bright red on the head of the male Western Tanager that made him think "cardinal." He assured me that it wasn't the finches, as he knew what they were and they were still coming to his feeder. He didn't know what a tanager was but the bird he was referring to didn't have any yellow on it. I thought that maybe I was fishing for the wrong clues. Perhaps the birds in question were crested and not necessarily red. "Do you know what Cedar Waxwings are?" I asked. He assured me that he knew Cedar Waxwings, but that they certainly didn't look like the birds he was talking about. I asked once more if he could describe them for me. He said, "They look exactly like the cardinal on the East coast except they aren't red and they don't have a crest." I'm afraid that I could be of no help, and to this day, I still don't know what "western cardinals" are.

I worked in the Biology Department at the University of Oregon for 30 years and each summer, I continue to teach the Field Ornithology course. (Even after retiring, I still return to teach this course each year.) Calls made to the department about birds were thus directed to me.

One May afternoon, I answered my phone and heard the department receptionist tell me that she was forwarding a call to me. The caller asked if I could help him identify a bird he had seen that afternoon at a park just north of Eugene. "It looked like a butterfly but when I got closer, I could see that it had feathers and don't tell me it was a hummingbird because it wasn't." With hummingbirds

Could someone have thought this Cedar Waxwing was a "Western Cardinal?"

Sphinx moths are large, colorful moths that have sometimes been confused for hummingbirds by people unfamiliar with them.

being ruled out, I thought perhaps he had seen a sphinx moth. These large, day-feeding moths are sometimes mistaken for hummingbirds as they hover over flowers from which they seek nectar. If you are close or if there are many of them, you can even hear the sound of their wings, just like you do with hummingbirds.

"No, no. It wasn't a moth and it didn't look like a hummingbird. I know what moths look like. This was a bird. It had feathers but it looked just like a butterfly. You know, those big orange and black butterflies. In fact my friend has a picture of one in a magazine and it looks just like that!"

Things began to click into place in my mind. That year, the April issue of *Natural History Magazine* included an article that was an April Fool's joke. There were several paintings of "little known birds." This included the Monarch Warbler[1]—an imaginary warbler whose wings looked like the monarch butterfly. As diplomatically as possible, I explained that the illustrations he had seen were in jest and not illustrations of real living birds and since we do have some monarchs in this area, maybe that is also a possibility. "Well, of course it didn't look exactly like the magazine but it did look like an orange and black butterfly except that it was a bird because it had feathers." I didn't know what else to say but, as it turned out, no more explanation was necessary as he had hung up on me. Immediately after that call, I walked over to the Biology Office in another building to check my mailbox. As I entered, a student worker was on the phone and asked the regular office staff, "Who do I forward a call about birds to?" Of course, they all immediately pointed at me. I took the phone only to discover it was the same man I had just been talking to and he was beginning his line of questions all over again. I patiently began to answer his questions once again. He started to explain more, then stopped in mid-sentence. "Hey, aren't you the same guy I was just talking to?" I said that I was. "Well... you don't know anything!" he said as he hung up.

---

1 You can see an illustration of the "Monarch Warbler" for yourself on page 53 of *A Field Guide to Little-known & Seldom-seen Birds of North America* by Ben, Cathryn and John Sill. 1988. Peachtree Publication, Ltd. This small book provides many fun and humorous accounts of fictional birds with marvelous illustrations by John Sill.

34

Do Birds Have a Sense of Smell?

Have you ever found a baby bird that has just fallen out of the nest? Maybe you see the nest in a small tree just overhead and it would be easy to return the baby to the nest. But you might have some hesitation because someone, perhaps even a teacher in your past, has told you that if you touch the nest, when the mother returns she will smell you and abandon the nest and her babies. What should you do?

The first thing that you should do upon finding a baby bird out of its nest is to return it as quickly as possible and step away to allow the parent to return. Maternal (and in many cases, paternal) instinct is very strong and the parent will try to protect that chick from harm, but if it is on the ground and unable to fly, it could easily fall prey to a lurking cat or other predator. You assistance may help save that young bird.

But will the parent smell you and be distressed?

First, the instinct to feed and protect babies is

Photo used with permission of Cascades Raptor Center.

The Turkey Vulture has a very keen sense of smell.

very strong in most birds and a mother will nearly always accept that chick you have returned to the nest. Secondly, ornithologists have long believed that most birds have little or no sense of smell, and therefore, the returning parent will not be able to detect your presence by odor.

The portion of the brain responsible for smell, the olfactory lobe, is very small in most birds. Therefore, it seemed logical to assume that birds have no sense of smell. With the exception of a few notable species, little work was done concerning smell in birds. The assumption is that there is little to test for if the olfactory lobe is poorly developed. But that turns out to be a faulty assumption and current research is demonstrating that some birds do have a good sense of smell.

Songbirds are generally not noted for having a well-developed sense of smell. The birds with the best sense of smell are: Kiwis in New Zealand, pelagic birds, like shearwaters and albatrosses, and Turkey Vultures.

Kiwis are odd birds in many respects. Unusual among birds, they use smell as their primary means of detecting prey—

earthworms. Their nostrils are at the tips of their beaks, not at the base as in most birds. They are nocturnal and feed by probing their long beak into the soft earth to smell for the presence of earthworms. When they remove their beak from the ground, they forcibly expel any dirt that is plugging their nostrils. Apparently you can hear them snorting during nightly foraging expeditions.

Albatrosses, shearwaters and their relatives spend nearly all of their lives at sea. Much of the ocean is a vast desert, but some areas are rich in life. The plankton which forms the base of the food chain has characteristic odors that are carried along the surface of the water. By following these scents, sea-going birds are able to locate the areas that are rich in life and food so as to find the fish and squid that they feed on.

The Turkey Vulture is probably the best North American example of a bird with a well-developed sense of smell. Since Turkey Vultures must consume dead animals for food, one way for them to find their food is by smell. That's where the mercaptans come into play... Turkey Vultures are especially good at finding sources of mercaptans, a class of chemicals we are familiar with for their unpleasant odors. Mercaptans smell like rotten eggs or rotten cabbage and we humans can smell it at very low levels. For these reasons, it is put into supplies of natural gas. Natural gas is methane, which is an odorless and colorless gas. If there were a leak in a gas line, we would have no way to detect it. However, by adding mercaptan, we can immediately tell if gas is leaking—it really stinks! Now, of course, it does the Turkey Vulture no good to follow natural gas, but mercaptans are useful for them to follow the scent of, because mercaptans are naturally present in the brain and parts of the abdomen of many animals. Upon death, these chemicals are released and more mercaptans may be formed. By following a scent trail of mercaptans, the Turkey Vulture is led to a likely carcass of a dead animal that it can feed upon. The carcass must be located visually, but the vulture is drawn to the area by the scent.

Only the Turkey Vulture and two other species in South America in the same genus have such a well-developed sense of smell. Vultures in Europe, Asia and Africa lack this sense of smell as do other American vultures, like condors and Black Vultures. However, Black Vultures have learned that by following Turkey Vultures, they can often be led to food they might otherwise miss.

So don't worry about re-placing that baby chick back into its nest. In all likelihood, the returning parent will never be able to smell you and the desire to care for its young will be strong enough for it to accept it back, assuming nothing else is wrong.

35

Winter: A Time for Breeding?

On cold, mid-winter days, as we watch birds seeking food, the long days and warm weather of the breeding season seem far away. Some people consider this a "slow" time when little bird activity can be seen, especially breeding behaviors. But if we venture outside and look more closely at the birds around us, we may discover that there are some exciting behaviors to watch, and some of these suggest breeding may be under way. In fact, breeding is very much on the minds of some birds in winter. We need only look to the nearest pond or river and view the ducks to see signs of the upcoming breeding season.

In most duck species, the male's brightest colors and patterns are those he is wearing in mid-winter. Actual mating will not take place until spring but many pre-breeding activities occur in fall and winter including the male's search for a mate. His bold colors are intended to attract the attention of prospective females, as it is the female who will ultimately select her mate.

As the seasons progress, an undefined signal indicates that it is time for the birds to leave their winter quarters and move to their breeding grounds, often hundreds of miles away. At some specified moment in the spring, the female determines it is time to leave, typically flying to the region in which she was hatched and raised. If a strong bond has formed between the two birds, the male follows his future mate to her breeding area, where a territory will be established and mating will soon occur.

Later in the breeding cycle, this early pair-bond breaks down and males abandon their mates. For some species, the bond is broken once the female begins the process of building a nest. In others, males may stay until eggs are laid or even until hatching begins. Without a bonded mate, some males return to the area in which they were raised, perhaps a long distance from their current breeding site. But many males stay in the breeding area seeking opportunistic matings with females with whom they share no pair-bond.

By mid-summer, the breeding season for most waterfowl (ducks and geese) has ended and you might notice that the male ducks no longer have their brightly-colored plumage. Instead, their feathers show cryptic patterns similar to those of females, because, until the next breeding cycle begins, bold colors and patterns are now a liability. Earlier, the need to attract a mate out-weighed the risk of attracting a predator. In female ducks, cryptic patterns help conceal them from danger as they sit on their eggs or raise their young. Unlike the majority of birds, ducks and other waterfowl molt all of their flight feathers at one time, rendering them flightless until the newly formed feathers fully emerge and are ready for flight. This seasonal flightlessness occurs in many water birds when the old feathers have become worn and weak, and are more easily broken. The old feathers are soon pushed out and shed as the new feathers grow and emerge. The stronger new feathers are better able to assist the bird in its upcoming journey to winter sites where the process of mate selection can begin anew.

Have you noticed that, unlike ducks, male geese look like

the females and have no brightly-colored patterns? Ducks must find a new mate each year and male ducks rarely help with raising the young. Conversely, in geese, males remain with their mates, often for life, and are diligent parents, helping to protect and care for their young. Thus, bright colors are not needed in geese. In fact, they are a liability, potentially attracting predators to the adults or the flightless goslings. Year-round, dull, protective coloration is much more beneficial to the overall well-being of geese.

If you look at birds closely throughout the seasons, you will discover that waterfowl are not the only birds to begin breeding activity in mid-winter. Owls, especially large owls, are either nesting or beginning pre-nesting activities then as well. In temperate regions of North America, you will find that by mid-January or February, Great Horned Owls are often on nests and beginning to incubate a clutch of eggs. (See Chapter 32 for more information about the Great Horned Owl.) The breeding season for many large owls is very long and the young owls will not fledge until late spring. If breeding did not begin until spring, the young owls would have less chance of survival, as they would not be leaving the nest until the end of summer when food supplies begin to diminish. By breeding from December through February, the young owls will leave their nest and learn how to hunt at the time of year when food is most abundant.

Smaller owls, like the Western Screech-Owl, have shorter incubation and brooding times and usually don't mate until early spring. However, many of these have already established a territory and pairs of birds can often be heard calling during the late evening and early pre-dawn hours at any time of year. If they are in an area where you live, and if you listen, you can hear the male and female call to each other. Whether it is the deep hoot of the Great Horned Owl or the soft, bouncing whistle of the Western Screech-Owl, you can tell male and female apart by voice without any special training. Listen carefully and you may hear two owls, one

The breeding period for Great Horned Owls, from mating to the fledging of young, is very long. Most Great Horned Owls in North America begin breeding in late December or early January.

with a voice that is pitched somewhat lower than the other. Male owls are smaller in size than females but their voice is lower in pitch. Without even seeing the birds, most of us can quickly learn to distinguish male from female. Western Screech-Owls are easily found in some of the wooded areas of many U.S. towns and suburbs.

There is another bird that you might not think of as being capable of breeding in the winter and that is a hummingbird. The Anna's Hummingbird is non-migratory, but has gradually expanded its range northward from California to as far north as southern British Columbia. Winters in western Oregon and Washington can often be quite cold for extended periods of time, yet these remarkable birds usually survive the cold winters. In California, these birds may begin breeding in January and even in the cold winter of the Willamette Valley of western Oregon, many Anna's Hummingbirds are on nests by mid-February. By the time most passerines are beginning to nest, the Anna's Hummingbirds have long since fledged. (See Chapter 8 for more about hummingbirds.)

Anna's Hummingbird, in a snow-covered bush.

All too often, we tend to equate spring with the season of breeding, renewal and re-growth. But nature is complex and varied, and, as just explained, some birds find winter to be an appropriate time to begin the breeding process. There is never an uninteresting time to observe the world in which we live.

36 Predators at Your Feeder

Millions of people across the country enjoy attracting birds to their yards. Planting bird-friendly plants is one method of bringing in birds but most people provide food in some type of feeder. Seed and hummingbird feeders are the most popular feeders but many enthusiasts also provide suet and make water available.

When you make an effort to attract birds, you will be rewarded with a variety of avian visitors. However, you need to be aware that you may occasionally attract some birds that you hadn't planned on. The very birds that you want to attract and enjoy seeing can occasionally be an attractant for some species of hawks that specialize in eating birds: Sharp-shinned Hawks and Cooper's Hawks. Both of these are woodland species and both primarily eat birds. You may not like them raiding your feeding area but they are usually infrequent visitors and the birds that they take are most often those that are slower and perhaps weaker.

Sharp-shinned Hawks are frequently seen hunting birds at back yard feeding stations.

Their preference for feeding on small birds has caused these raptors to be severely persecuted in the past. So many Sharp-shinned Hawks were shot in the early 20th Century that some populations fell into serious decline. Even prominent ornithologists of the time thought it was acceptable to kill these hawks. In *Birds of Oregon* (1940), the authors, Ira Gabrielson and Stanley Jewett, say that the Sharp-shinned *"...is undoubtedly the most destructive hawk in the State."* The eggs of these birds were also highly sought after by early egg-collectors. Fortunately, hawks (and their eggs, feathers and nests) are protected under state and federal laws today and attitudes have largely changed to allow us to more fully appreciate all birds of prey. However, rehabilitation centers still see too many raptors come in because of gunshot wounds.

The Sharp-shinned Hawk is one of the smallest day-hunting raptors in North America, along with the American Kestrel. Which is smaller depends on whether you are looking at a male or female (female raptors are always larger than males) and in what part of the North America the birds occur. The Sharp-shinned Hawk also shows the greatest difference in size between male and female. In nearly all raptors, females are larger and heavier than males. The difference is most pronounced in the Sharp-shinned Hawk where males weigh less than 60% of the weight of females. This is also a highly migratory bird, although the species is often found year-round in many parts of the country.

Western Oregon is one place where Sharp-shinned Hawks are present throughout the year but the individuals found here in the winter may not be the same ones that breed here in the summer. Without marking individual birds, we will never know for sure, but we do know that many of the hawks that breed in Alaska and Canada spend the winter in Washington and Oregon. We also know that some juvenile females that were raised in the Pacific Northwest have been found as far south as Costa Rica by the winter after they were hatched. In the East, there is some evidence that some Sharp-shinned Hawks are not migrating, and instead, are choosing to stay where many feeders attract small birds.

While these birds may sometimes be found hunting in our urban areas, when it is time to nest, they look for stands of conifers. The nest is a large collection of twigs constructed by the female with materials gathered by both sexes. Once the female is on eggs (May to June in Oregon), the male will bring all of her food and all of the food for their young once they have hatched. When he brings food to the nest, the female takes it from him and tears it apart to feed the young. If the male then remains at the nest very long, the female often drives him away while she feeds the babies. When the young are ready to fledge, both adults will hunt and bring food.

Another bird-eating hawk that frequents bird feeders is the Cooper's Hawk. This hawk is closer to being crow-sized but in many respects, it looks similar to the Sharp-shinned Hawk. As in Sharp-shinned Hawks, females are considerably larger than males and like "Sharpies," conifers are the preferred nesting sites. But, Cooper's Hawks seem to be less particular about nest sites and many birds are now found in urban wood-lots across the country. Unlike Sharp-shinned Hawks, in Cooper's Hawks, it is the male that gathers most of the nest material and builds the nest. He even shares some incubation duties, spending 10-20 minutes on the eggs two or three times a day. Cooper's Hawks take some small birds but they prefer medium-sized birds, such as jays and doves. When near a stream, they sometimes take prey to the water's edge and hold it under water until it drowns. This is a quick method of killing prey and minimizes a possible struggle that could harm the hawk. Cooper's Hawks are also much more likely to capture squirrels or other small mammals than are Sharp-shinned Hawks.

Both Sharp-shinned and Cooper's Hawks have short, rounded wings and long tails. This shape gives them great maneuverability to chase birds through dense woody thickets. Even so, these hawks are sometimes injured when they hit a branch with their wings as they are intently pursuing prey. Examination of many wild birds shows evidence of healing from such wounds.

Since these two species look so much alike, I am often asked how to tell them apart. Many field guides tell you to look at the tail as the birds fly. Sharp-shinned Hawks have tails with feathers that are nearly of equal length, giving a square appearance to the end of the tail. The outer tail feathers of Cooper's Hawks are shorter which gives a rounded appearance to the tail. Unfortunately, this is not an entirely reliable feature to use. It depends on how widely the tail is spread, the condition of the feathers, and what stage of molt the bird is in.

There is, however, one method that is usually very reliable but not often given in field guides. When the bird is in flight, draw an imaginary line between the wrists (the bend in the wing) of the out-stretched wings. If the head is mostly tucked back behind this line, the bird is a Sharp-shinned Hawk. If the head looks large and extends out beyond this imaginary line, it is a Cooper's Hawk. I use one trick that might help you remember this difference: If the head is large and protrudes, it is like a barrel. Since a cooper is a barrel-maker, this must be a Cooper's Hawk. (The bird was really named for William C. Cooper [1798–1864], an American Zoologist and a founder of the New York Lyceum of Natural History.)

Whichever species it is, you may be initially disturbed if one swoops through your yard and takes a bird from your feeding station. But, if you can get past your initial distress, you may also be thrilled with the power and efficiency of these beautiful hawks. Enjoy, rather than discourage their presence. You will find that they take very few birds from your feeder and do so only infrequently. You may also find yourself feeling admiration and joy when seeing one of these magnificent creatures closer than usual.

37 The Importance of Vision

Many people enjoy gardening and landscaping their yards in ways that are restful and beautiful to the human eye...as well as to attract birds. The visual makeup of your yard is what will first attract birds. Fragrant flowers are pleasant to us and may attract insect pollinators, but most birds have little or no sense of smell, so fragrance is of no importance to them. The kinds of food and shelter that you provide are important, but how do the birds first find them?

As birds move from place to place seeking food and shelter, they are examining their opportunities in much the same way that we would—by looking at the world around them. Vision is of critical importance to birds and birds' vision is among the best of any animals on Earth.

The visual world is very important to most of us as well. We derive joy from the variety of colors and shapes that we perceive and they provide vital clues and much information about the world in which we live. But our vision is

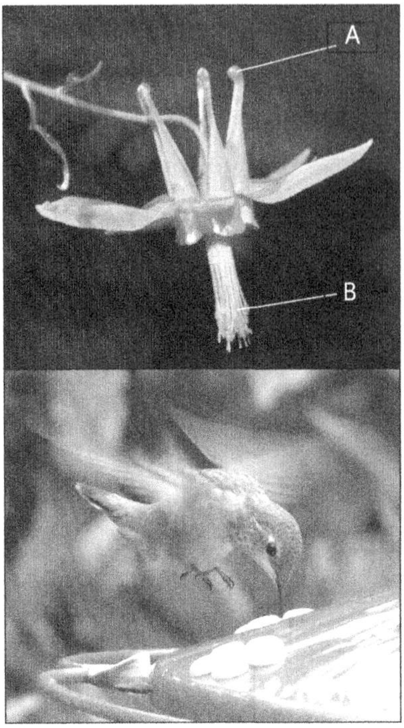

The columbine flower (top) has evolved to attract hummingbirds, like the Rufous Hummingbird (bottom). The red color is easily seen and the flower hangs upside-down which discourages butterflies, but not hummingbirds. When the hummingbird reaches for the nectar at the top of each tube (A), its head is dusted with pollen from anthers hanging from the bottom of the flower (B). The hummingbird can reach its treat and the flower benefits by having the hummingbird carry pollen to another columbine bloom.

impoverished in comparison to that of birds. We can only guess at what birds actually see because we ourselves cannot see all that they can.

Our retinas have cells called cones, each of which is sensitive to one of three primary colors: red, green or blue. From a mixture of those three colors our brain creates the multitude of colors that we know. We see better than most other mammals because, aside from humans and other primates, most mammals' eyes are sensitive to only two primary colors: blue and red. However, the visual abilities of birds exceeds us all, for birds see not only red, green and blue, but they also see ultraviolet, a region of light beyond our ability to see. Additionally, the cones in a bird's retina have an additional structure that we lack: each cone has an oil droplet associated with it. Some of these droplets filter particular wavelengths of light, and others enhance and provide contrast to other wavelengths of light. The result is an incredible visual system able to distinguish more kinds of light and in greater detail, and we can only guess and imagine what they see.

In addition to color, the cones are also responsible for

sharpness. The more cones per unit area, the greater the visual acuity or ability to resolve fine detail. If an image is not sharp, magnifying it only produces a larger blurry image. Resolving the image in finer detail is what is important and it is the cones that allow animals to see detail.

In the fovea, or center portion of the retina where you can see the most clearly, humans have approximately 200,000 cones per square millimeter. Many raptors, swifts and flycatchers have over one million cones per millimeter squared. Birds can see detail and color far beyond that capable of the human eye.

These facts are all very interesting to a biologist, but how might this be a consideration for you in your garden?

You may know that hummingbirds are attracted to red flowers, but like other birds, hummingbirds see much more than just red. However, red provides a very strong contrast to green and this contrast is enhanced by the bird's visual system. Most foliage is green, so a red flower becomes very easy to see even at a distance. Blue flowers do not contrast nearly so well with the green foliage and are less attractive at a distance. So, the initial attraction is to the spot of red. Upon closer investigation, the hummingbird sees the flower more clearly and decides whether it is one to forage from, and then once attracted to a region, other flowers may also be of interest. Larkspur, for example, is blue and is attractive to bumblebees, but hummingbirds will also feed on larkspur if they find it blooming in a region in which they are foraging. So, in conclusion, planting red-flowering plants with hanging flowers, preferably species native to your area, helps attract hummingbirds to your yard.

Ultraviolet light also reflects differently from dead or dying foliage than it does from healthy, living foliage or other subjects. A brown mouse in dry grass could be very difficult for you to see, but to a bird, the brown fur of a living mouse will look very different than the surrounding dry grass.

Some fruits look different in ultraviolet light when ripe, allowing birds such as this Cedar Waxwing, to know when is the best time to eat them.

At least one raptor, the Eurasian Kestrel, uses ultraviolet light to help it hunt. Mammals mark their territory by spraying urine and a compound in the urine, urea, glows in ultraviolet light. Mammals smell the scent marks of others but cannot see the glow caused by the ultraviolet. The kestrel, however, ***can*** see the glowing urea and simply needs to look for the brightest glowing spot to find the freshest scent mark in order to maximize its chance of finding its rodent prey. Our American Kestrel can also see ultraviolet light but it is not known if it also uses this visual aid while hunting.

Ultraviolet light helps birds in other ways, too. They are better able to see some fruits to assess whether they are ready for eating, because in ultraviolet light, many ripe fruits look different than they do when they are not fully ripe.

Patterns on insects, visible only in ultraviolet light, may also assist birds in finding these insects.

We also are beginning to learn that the plumage of many birds looks different in ultraviolet light. For example, male and female Cedar Waxwings may look alike to our eyes, but there is apparently a difference that can be seen in ultraviolet light which the birds can see, and we cannot. European Starlings are another species where males and females look very much alike to us, but during the breeding season, the throat of the male glows in ultraviolet light. In fact, in the species where male and female look alike to our eyes, over 70% of those show differences in ultraviolet light. How the birds use these differences is still a matter of study.

Our gardens may look beautiful to our eyes and be a sea of green splashed with color throughout. But birds examine our gardens with very different eyes. They see many more colors than we do and those colors give them clues about the potential for food. A good variety, especially of native vegetation as much as possible, will attract the largest variety of birds as they seek shelter and food. And as they fly overhead, they will first be drawn to your garden by what they see.

38

Angel Wing

Finally! The sun peeks out, so you head to the park. You watch as a family scatters bread and popcorn for the ducks and geese and the birds swim over to eat the handouts. The children squeal with delight, the parents smile and the ducks and geese enjoy an easy meal. It seems an idyllic scene with all parties going away happy...or is there a hidden danger?

There *is* a hidden problem in this story, however. Despite the family's good intentions, they are unknowingly endangering the health of the ducks and geese, so eager to take the handouts offered. Human-fed waterfowl are put at risk of developing a condition known as "angel wing." You have probably seen birds with this condition without knowing the cause. The tips of their wings point outward, away from the body, as if broken, and, sadly, these birds cannot fly.

Angel wing affects geese, and to a lesser extent,

Sadly, this goose will never be able to fly. It was fed bread or other human food when it was a chick, thus condemning it to a life without flight and at higher risk of danger from dogs and those who enjoy harming poor creatures that can't escape. Enjoy the birds in parks, but DO NOT feed them! You can cause them great suffering and early death since they'll never be able to fly.

ducks and swans, but it is a problem completely unknown in totally wild populations of waterfowl. It is found only where people feed birds, mostly in parks where waterfowl congregate. ***If you care about the birds' health, don't feed them***; just enjoy watching them.

Much is unknown about angel wing, but it only occurs as young birds are growing. A diet high in carbohydrates and protein is the cause of the problem and this is precisely what some human foods (popcorn, bread, etc.) contain. Large amounts of these foods raise the birds' sugar and protein levels and decrease levels of manganese and vitamins D and E. The imbalance of these dietary constituents causes the feathers of young birds to grow too fast. Thus, the weight of the feathers in the wings becomes greater than the birds' developing wrist bones can support and the bones twist outward. As the bones grow and solidify, the wings' twisted position becomes permanent, condemning the bird to a flightless existence. Both wings are often affected, but sometimes only the left wing is twisted. For reasons we don't understand, rarely is only the right wing affected. ***By not feeding the birds foods unnatural to them, angel wing can be completely prevented.***

Cracked corn can usually be safely use and not cause Angel wing, but there are other dangers to consider from feeding birds at ponds and lakesides in parks. The birds will become accustomed to people and have less natural fear. In parks this is usually not a problem, but if these same birds show no fear at other locales, they may be putting themselves at greater risk from people who may do them harm. At parks, there are often many people who bring their dogs with them and some dogs will chase the birds and may catch and injure or kill some of them. Dogs may be a greater threat in parks where waterfowl are encouraged to congregate because they are fed by well-meaning people. The large number of waterfowl that can gather at parks can potentially present a health problem for children. A large number of ducks and geese can create quite a mess with their droppings. Children love to play around the water's edge and they can get these droppings on their bare feet and hands, and young children often put their hands in their mouths, increasing the possibility of fecal contamination.

Where large numbers of waterfowl gather around still water, like ponds in parks, there is another potential risk to the birds—botulism. This disease is caused by the bacterium *Clostridium botulinum* and seven different stains of this bacterium have been identified. It is sometimes carried in the droppings of infected birds and in stagnant water conditions with high levels of protein, the bacteria can proliferate and cause widespread illness, lesions, paralysis, and often death, in the waterfowl gathered together in these waters. This can be a great health hazard to the birds but it is not a health

Another view of a Canada Goose with Angel Wing, a disease caused by people feeding them bread when they were chicks. Since these geese can't fly, they will leave their droppings where children might access them.

hazard to people. The strain of botulism that infects birds is a different strain than the botulism bacterium that causes illness in humans. Thorough cooking will destroy the harmful toxin produced by this bacterium, but it is simply safer to avoid contaminated areas. If many dead birds are found, avoid the water and call health and wildlife officials.

If feeding geese and ducks at the park is sometimes harmful, what about feeding birds in our back yards? In your yard, the situation is quite different than feeding popcorn and bread to ducks and geese. In the park, the young, rapidly-growing birds are put at risk but in yards, most birds at feeders are adults, or fully-fledged birds whose wings are formed. The seed in your feeders will not be fed to young birds in nests since most songbirds feed only insects to their young. Seeds sold as bird feed are either natural, or similar in properties, to seeds the birds eat in the wild, and angel wing is not known to occur in songbirds.

There are many misconceptions about feeding birds. If you have heard that birds may not migrate if you don't take down your feeders, be assured that this is completely false. In migratory birds, the urge to migrate is too strong to be overcome by the presence of feeders, and food we supply accounts for only a small portion of any bird's diet. There are some risks to feeding birds in your yard, but these are minimal and preventable with a little effort on your part.

What are these risks? Some bacterial and fungal diseases can easily be spread among birds congregating at feeding stations. But, you can prevent this by keeping feeders clean and free from debris. If you see any sign of mold on your feeders or the ground, clean it away immediately. Feeders can be cleaned with a weak bleach solution, but if you do this, be certain to completely rinse away all traces of bleach or detergent residues. If you live in an area with wet winters, it is advisable to cover your feeders so that the seed cannot get wet, causing mold to grow on the feeder or seeds. A variety of plastic domes are available that work well.

Another risk is cats, both feral and pet cats. Be sure to place your feeders in such a place that cats do not have the opportunity to reach or jump up to catch birds coming to your feeder, at least 6 feet high. (Some discussion of the dangers of cats is found in Chapter 7.)

Feeding birds can bring much pleasure but we need to do it in a manner that causes no harm to the birds or to other animals. Only a small effort is needed to protect the birds you care about feeding. In your yard, a bird feeder is a wonderful way to see birds, but away from home, enjoy observing the ducks and geese in your local parks, but please do NOT bring bread, popcorn or other human food to feed waterfowl in our parks. It is not a kindness to feed birds when doing so can jeopardize their health or potentially cause deformities such as angel wing.

39

Thistle-bird

Throughout history, people have used some very colorful names for the birds they see. Wild Canary, Lettuce Bird, Thistle-bird, Beet-bird, Arkansas Goldfinch and Willow Goldfinch are a few of the names that have been used for the American Goldfinch. Many readers may be acquainted with this bird of open country, but most know it only by the bright yellow plumage worn by the male in summer. The females are duller and less often recognized. During the winter, both male and female are a dull yellow-green in color and can be easily overlooked, despite regularly visiting feeders.

The American Goldfinch breeds later than any other species of bird in North America, often not until mid-July or even early August. When other birds are caring for youngsters or starting a second brood, American Goldfinches are just beginning to build their nests. Why so late? They are waiting for a good supply of seeds, especially seeds with downy tufts, such as thistle seeds (thus the name "Thistle-bird.") These soft downy plant parts are used to line

The American Goldfinch feeds its young small seeds instead of insects, as most birds do.

the nest and the small seeds are fed to the young once they hatch.

The majority of seed-eating birds will feed insects to their newly hatched young because insects are softer than seeds, they are protein-rich and are much easier to digest. But, contrary to this norm, the American Goldfinch feeds small seeds to its young, starting with their first meal. The adults may crack the seed coat or even partially digest a few, but mostly the young must be able to eat whole small seeds. Such heavy dependence upon seed-eating is partly what pushes the goldfinch to nest so late.

But, nesting late also means that there is usually only time to raise one brood of youngsters. In a few instances (less than 15% of the time) American Goldfinches are known to have second broods. In order to have time to raise two broods, the female must abandon her first set of young shortly after

The non-breeding plumage of the American Goldfinch lacks the bright yellow of the breeding male.

they have hatched. Her mate must take over all care of the young while the female seeks another mate with which to raise her second brood. The first male is well-conditioned to bringing food as he has brought his mate all of her food once she started sitting on eggs and he continues to bring food to the young once they hatch. If his female has left to start another brood, he will continue to feed the young, but while he searches for food, he cannot provide the full attention a female at the nest would give, exposing the young to potential danger while he is away.

Most American Goldfinches pair up for mating while still in winter feeding flocks. Sometimes, those pair-bonds break down and new mates are established prior to mating. This mate change is initiated by the female as she chooses her prospective mate for the season. Her first choice is made during the winter months when males are still a dull yellow. In the spring, goldfinches undergo a full body molt, replacing all but the wing and tail feathers. After this molt, the males will be bright yellow, and the brighter the yellow, the more attractive a male is to females. If the female's first choice of mate is duller than other males, she may re-select a more brightly-colored male.

The yellow color comes from plant pigments (carotenoids) in the bird's diet, so a brighter yellow indicates a healthier male with a more carotenoid-rich diet. Carotenoids are important not only for feather color, but they help boost the bird's immune system. A bird fighting an infection during its time of molt will be less colorful because the carotenoids are used by the immune system rather than being stored and later deposited in the newly forming feathers.

In the northern portions of their range (northern U.S. and southern Canada), American Goldfinches are usually migratory, but they are present throughout the year in much of their range, including where I live in Oregon. As winter approaches, they become acclimatized to cold and can survive long periods of very cold weather.

They are not overly attracted to bird feeders during the breeding season when plenty of wild food is available, but during the winter, they can often be found at feeders. Black oil sunflower is one of the best foods to provide and is eagerly taken by goldfinches. Nyjer seed (formerly called "thistle" seed) is also attractive to goldfinches but is expensive and sometimes may sit for long periods before being discovered by the birds.

Once the winter plumage is present, some people think that American Goldfinches have departed, not realizing that their bright yellow feathers are replaced in the fall by others that are dull yellow-green. They are also sometimes confused with another species of goldfinch, the Lesser Goldfinch, which is just as easily attracted to feeders during the winter, although often in fewer numbers. The two species often forage together. In the breeding season, the male Lesser Goldfinches are not as bright yellow as American Goldfinches and are darker over the back and head. During the winter, they can be easily distinguished in flight by looking at the rump. The Lesser Goldfinch has a dark rump in contrast to the white rump of the American Goldfinch, and in all plumages, the Lesser Goldfinch shows more white in the wing.

The Lesser Goldfinch, also attracted to sunflower seeds, is less common in many localities. It has more white in the wing and the male has an all dark cap, not just a black forehead as does the male American Goldfinch.

Unfortunately, American Goldfinches fall victim to cats more frequently than many other birds. Although they generally avoid direct contact with the ground, they often feed on plants that are low to the ground, especially when plants like thistles are producing large quantities of seed. They are alert to avian predators, but for some reason, they often tend to ignore mammals, which gives cats an opportunity for close approach. So if your feeders are attracting goldfinches, be sure that they are unreachable by cats. If you live in a heavily wooded area, you will likely not get goldfinches at yard feeders, but open regions and suburban neighborhoods are likely places to attract goldfinches. You will see them mixed with other birds such as House Finches and Pine Siskins all enjoying a feast of sunflower seeds that you have provided.

40

Some Non-natives are Here to Stay

Probably everyone has seen a House Sparrow, perhaps at a bird feeder. Even those who care nothing about birds likely have crossed paths with this bird in a grocery store parking lot or similar location. If you pass through agricultural lands, you will find them near out-buildings and in the fields. They closely associate with humans and they avoid wilder places.

You may know the House Sparrow by its old name, English Sparrow. It also has many colloquial names: Bull Sparrow, Gamin, Hoodlum, Parasite, Town Sparrow, Tramp, and Domestic Sparrow. You'll notice that many of these names are unkind and that's because this small bird is not well-liked by most in America who know it.

The House Sparrow is a native of Europe where it is also found in close association with humans. It was deliberately introduced in this country in an attempt to control inch-

Male House Sparrow showing his large black bib. Males with the largest bibs are the most dominant.

worms that were damaging shade trees planted along city streets. First released in Brooklyn in 1851, they were initially well cared for and protected. It didn't take long, though, for people to realize that the damage caused by House Sparrows greatly out-weighed any potential good. In fact, they did little to bring the inch-worm pests under control and their reproductive potential was astounding. A government report issued in 1889 showed that in just ten years a single pair of House Sparrows and their offspring could result in 275,716,983,689 more sparrows. This didn't happen, but they did quickly become a dominant bird in city landscapes along the Atlantic Seaboard. As horses pulled wagons through city streets, the birds followed, feeding on spilled grain. As the birds' population grew, they spread into agricultural areas, where they found more grain to feed upon.

House Sparrows quickly spread across the country to the

western states and they were first reported in Portland, Oregon in 1889. Soon they were found throughout the state in cities and agricultural regions. A report in 1921 found them abundant in the town of Burns (in southeast Oregon) and in surrounding ranch lands. It is believed that many of the birds reached the West coast by becoming trapped in railroad grain cars. Then, as motor vehicles increased and replaced horses in cities, the population of House Sparrows declined from past high numbers, but certainly did not disappear as a city and farmyard bird.

House Sparrows have been introduced nearly worldwide and in most cases, the results have been similar, seriously impacting agriculture and causing some disruption of native cavity-nesting birds. Some interesting changes have occurred in different populations, however. In their native Europe, House Sparrows are non-migratory as is true here in North America. However, in Central Asia, where they were also introduced, House Sparrows have developed interesting migratory patterns, with some birds moving to different geographic regions and some wintering at low elevation and moving to over 6,000 ft. to breed.

Here in North America, House Sparrows in the far north are generally larger than their southern relatives, and birds of moist environments (as found here in the Pacific Northwest) tend to be darker than birds of arid regions. (These patterns are true of other bird species and mammals as well.)

Our native sparrows, such as Song Sparrows, are not closely related to House Sparrows. The House Sparrow (and its cousin the Eurasian Tree Sparrow) are most closely related to weaver finches of Africa, known for their elaborately woven nests. But nest-building is not a strong attribute of House Sparrows. They select a cavity in which to build a somewhat untidy collection of dried vegetation that often contains scraps of paper or pieces of string. Some fresh green vegetation, often mustards or mints, are also usually included. House Sparrows are very aggressive and in their search for cavities, they will often drive out native cavity-nesting birds. Bluebirds and swallows are frequently driven out and, if young are present, those young may be killed. The potential for damage to native species is high because House Sparrows can have as many as 4 successful broods per year and they are known to live as long as 13 years.

Because they are non-native and can cause damage to human resources and to native birds, most of us tend to focus only on the negative aspects of this species. They are here to stay and we can benefit by understanding more about their biology and behaviors. By overcoming our bias, we can also observe some interesting behaviors. For example, look closely at a group of House Sparrows. Male and female are easily distinguished because males are more colorful and have a black bib. But males are not all identical. Some have larger black bibs than others. If you observe them closely, you might notice that males with larger black bibs are also the most aggressive. The larger black bib also attracts the attention of females who prefer these males over those with less black.

Their numbers, and the lack of legal protection, makes House Sparrows ideal subjects for many scientific studies. Along with another non-native, the European Starling, much of what we know about songbird physiology and biology comes from studies done with these birds. Zoologists who study these birds have learned information that aids in the understanding of other birds, and can be helpful to biologists in Europe, too. For as much as we consider the House Sparrow a pest in this country, there are serious concerns for this bird in portions of Europe where it is native and now in serious decline. Why a species, especially one so closely tied to human habitations, is rapidly decreasing and becoming threatened, is a problem that we need to understand, for the benefit of the species itself and for the implications the decline may have in terms potential threats to other species.

About the Author & Illustrator

Dan and Barbara Gleason are a natural science writer-photographer and artist-illustrator who love working, birding and exploring the outdoors. Their first book was Birds! From the Inside Out, first published in 1999 and now in its fourth edition. Dan writes a monthly column 'Backyard Habitats,' for the *Register-Guard*, the major newspaper for Eugene, Oregon.

Dan worked in the Biology Department at the University of Oregon until retiring and he continues to teach Field Ornithology each summer. He has taught workshops at the Malheur Field Station, Sitka Center for Art & Ecology, and other centers, and he has developed and taught docent trainings on natural history and bird biology. He is a popular invited speaker at bird conferences and festivals, and for Audubon chapters and bird- and nature-oriented organizations.

Dan served as a director on the board of directors of the Lane County Audubon Society and Oregon Field Ornithologists, and he volunteers at Cascades Raptor Center, providing instruction about birds to staff, volunteers and the public. He enjoys retirement by writing, photographing and exploring the outdoors, as well as restoring old photographs.

Barbara is a colored pencil artist, natural science illustrator and graphic designer who manages BGleason Design & Illustration, and CraneDance Publications, their book-publishing wing. She provides design, illustration and consulting services to clients that range from individual authors to numerous companies and government agencies.

A member of the Guild of Natural Science Illustrators and the Colored Pencil Society of America, Barbara has also been active in a number of business and art groups. She served on the board of directors of Lane County Audubon Society and is the corporate image director for the Colored Pencil Society of America.

Barbara paints birds, wildlife and landscapes and helps maintain a bevy of wild bird feeders in order to study her subjects better.

Together, Barbara and Dan have taught Nordic Walking and they enjoy many activities out in nature. They live in Eugene, Oregon, with two indoor-dwelling bird-watching cats who enjoy the turkeys, squirrels and deer from their enclosed "catio."

Barbara and Dan, along with their friend, Becky Uhler, a fellow artist-birder, will open a new Wild Birds Unlimited store in Eugene in early 2013. Bringing such a store to their community fits well with their passion for serving that community by providing a bird feeding and education resource for those who love birds. Their mission for the store is to help bring people and nature together, while improving the welfare of the birds they love and the connection with nature for the people who frequent their store.

The Gleasons can be reached by email at bdgleason@comcast.net, by mail at P.O. Box 50535, Eugene OR 97405, or by phone at (541) 345-3974, West coast time, between the hours of 10 am and 6 pm. They may also often be found during business hours at the Wild Birds Unlimited store in Eugene, Oregon, (once it opens in Spring 2013) that they own with thier freind Becky Uhler.

Made in the USA
Monee, IL
13 September 2026

62981002R00154